VIETNAM
Its Impact On My World

VIETNAM
Its Impact On My World

Stephen E. Manthey
Brian Manthey

Vietnam: Its Impact On My World

Copyright © 2022 by Brian Manthey

All rights reserved. No part of this publication may be reproduced, distributed, or transmitted in any form or by any means, including photocopying, recording, or other electronic or mechanical methods, without the prior written permission of the author, except in the case of brief quotations embodied in critical reviews and certain other noncommercial uses permitted by copyright law.

Jones Media Publishing
10645 N. Tatum Blvd. Ste. 200-166
Phoenix, AZ 85028
www.JonesMediaPublishing.com

Disclaimer:
The author strives to be as accurate and complete as possible in the creation of this book, notwithstanding the fact that the author does not warrant or represent at any time that the contents within are accurate due to the rapidly changing nature of the Internet.

While all attempts have been made to verify information provided in this publication, the Author and the Publisher assume no responsibility and are not liable for errors, omissions, or contrary interpretation of the subject matter herein. The Author and Publisher hereby disclaim any liability, loss or damage incurred as a result of the application and utilization, whether directly or indirectly, of any information, suggestion, advice, or procedure in this book. Any perceived slights of specific persons, peoples, or organizations are unintentional.

In practical advice books, like anything else in life, there are no guarantees of income made. Readers are cautioned to rely on their own judgment about their individual circumstances to act accordingly. Readers are responsible for their own actions, choices, and results. This book is not intended for use as a source of legal, business, accounting or financial advice. All readers are advised to seek the services of competent professionals in legal, business, accounting, and finance field.

Author: Stephen E. Manthey
Editor: Brian Manthey and copyright holder of the contents

Printed in the United States of America

ISBN: 978-1-948382-46-5 paperback

TABLE OF CONTENTS

Introduction vii

Chapter 1: A Gift From My Dad. 1

Chapter 2: A Gift From Grandpa. 5

Chapter 3: Death Fear 13

Chapter 4: Innocence 19

Chapter 5: Dreams 27

Chapter 6: Therapy. 31

Chapter 7: Leadership 37

Chapter 8: Children 47

Chapter 9: The Survivalist 55

Chapter 10: Crazy People. 65

Chapter 11: Christmas 79

Chapter 12: Fear 87

Chapter 13: Animals In War 99

Chapter 14: Unexpected. 107

Chapter 15: Best Friends 119

Chapter 16: Coming And Going. 133

Afterword. 147

INTRODUCTION

One year, one month, one day. 396 days One 47th of my life. The impact of that year on my life can not be measured in time. All the years after that year have been impacted and changed by that one year.

Yet, as I watch other veterans, I realize the impact on them is different then it is on me. I realize the 18 to 25 years before that year not only impacted how I related to the one year in war but also has impacted the years that followed.

There is an interweaving of events that play upon each other. This interweaving goes beyond myself. As I relate in one of the following stories, my father was an avid pheasant hunter. He pushed hard to get me involved in the sport. This event probably saved my life.

I have a brother. He is three years younger than me. As we grew up, we shared many experiences. We both spent our summers working on my uncle's farm. We both survived the divorce of my parents. We both went to the same schools and had many of the same friends. Admittedly, we were different. Yet, we are more alike then other people. My brother has held three jobs in his life. I joke that the longest job I ever held was the time I spent in the army. It is not far from the truth. Why?

Is the answer to this question in the major differences between my brother and me? Is the answer. One year, one month, one day of my life?

This series of stories is the result of my search for this answer.

Yet, there is still another reason I wrote this document. For over 20 years, I have kept these experiences in my mind. When I began writing, I was surprised how current my memories felt. It is because I relive Vietnam every day of my life.

When I wrote my first chapter, I was relieved that the memory faded. It is as if I do not have to keep my memories alive any more. They are locked into the memory of the computer. And at last, I am free. As free as I will let myself become.

When I saw my best friends name on the Vietnam memorial in Washington DC, I cried.

I cried because I was not the only one trying to keep the memories of my friends alive. This does not change my feeling of responsibility. In my heart, I am the one person who must remember.

Stephen Edward Manthey, American

CHAPTER 1

A GIFT FROM MY DAD

January 31, 1969, midnight. I was in Vietnam and the biggest battle of my time in the war was about to begin. The NVA had tried to move into the town we were guarding and we had redirected their movements. A helicopter had just spotted an estimated 300 enemy troops in the rice paddies.

I was on the Armored Personal Carrier (track) covering the left flank. There were five tracks in our platoon with five solders in each track. Our weapons were a fifty caliber machine gun and two M60 machine guns. I was the driver.

As soon as we spotted the enemy in the paddies we formed a line and opened up all 15 machine guns. Suddenly the left machine gun the one behind the driver, jambed. Ahead was a four foot high dike. It was the rainy season and the mud in the rice paddies was thick. It was impossible to climb the dike. We stopped.

Immediately, about 15 feet away, an enemy soldier stood up. He was holding a rocket launcher. A flair popped overhead and his eyes met mine.

Vietnam

Without thinking I grabbed the M16 kept next to my seat and raised it in one fluid motion. As it came up I cocked it and released the safety. The rifle tucked into my shoulder and I looked down the sights into the terrified eyes of the enemy.

Years earlier, at the age of 13, I was trained to shoot a shotgun by my father. I was an excellent shot and could easily out shoot everyone except my brother and my dad. We would spend hours at the firing range shooting clay pigeons. I loved shooting the 12 gauge and looked forward to those times.

Hunting pheasant was a different story. My father believed that shooting clay pigeons was practice for hunting and the only reason to shoot was to hunt. He would take me to Eastern Washington every weekend during hunting season and we would hunt. Dad usually shot his limit. I never brought home anything.

In my memory it seemed that we were always discussing why I missed. My mother would blame herself, saying that she had spent too much time teaching me that killing was a sin. My father just couldn't understand how I could be such a great shot and always miss the birds.

The last day of hunting season, my father watched me instead of hunting. This was a great sacrifice for him because he loved to hunt.

As we walked along the fields with the dog in front of us, a pheasant jumped up under my feet. It was so close I felt the wings hit the sides of my legs as it took off. I almost dropped my shot gun. I was so flustered I didn't even aim. I just fired. My father said that he understood. I had a case of "buck fever." When a bird would take off, I froze. Finally, I would act but it was too late.

Seven years later, I was in Vietnam. I was facing the enemy and was about to fire. The man I was facing was scared. He was frozen and unable to react. As my rifle came to my shoulder, he suddenly began to raise his weapon. I fired. This took an instant. Less then a second. It all happens so fast you can't think. You rely on instincts and training.

Hector, my hunting dog died years ago. Lloyd also died years ago of cancer and my father is also gone. My fathers patients and understanding had trained me for Something that later kept me alive. I didn't tell my dad that he had saved my life. It would seem so strange to tell him that he had not only given me life but he also kept me alive. And what could I say? Thank you seems so much less then what I feel.

That year before hunting season we went to the range as usual. This year Dad had a different approach. I would start at my Dads side and walk toward the low concrete block house where the clay pigeons were released. Instead of yelling "pull' and being ready, my father would release the flying discs when he felt like it. I would jump a mile and always miss. We practiced and practiced. Looking back on it I am amazed at my Dads patients. After many many attempts and becoming very discouraged, I started again. I walked right up to the house and he still hadn't released the target. I stopped and waited. He still didn't let it fly. I turned and said "Daaaaaad!!" just as he released the clay pigeon. I turned and fired and nailed it.

From that time on I began to hit more and more until finally it didn't matter any more. My dad said I was ready for the hunt.

Our first hunt of the season was with LLoyd Severts, and our hunting dog, a Britney Spaniel named Hector. It was a wonderful fall morning and we were hunting a draw. The dog and Lloyd were in the heavy brush at the bottom of the draw. Lloyd was a lot of fun and one of my Dad's best friends. He always brought a new shot gun that was guaranteed to hit every time. He usually missed. My dad was on the left side and I was on the right. Dad kept saying that the dog was so excited that he was sure there was a rooster in this Draw. Dad always said that. I was enjoying the feeling of the sun and the smell of the sage brush. I was also glad that Lloyd was in the heavy brush. That was usually my job.

Suddenly a male pheasant exploded from the grass next to me. At first I heard the loud noise. Then I saw the red colors flash in the sun and saw his long tail. He flew from my left to my right. With one fluid motion I brought the 12 gauge up and fired. The shot was right on target and the pheasant came down on a puff of feathers. I couldn't believe it. I ran to the bird and picked him up. Then I turned and lifted him high over my head. My Dad was looking at me with the biggest smile I had ever seen. He had seen the whole thing.

That was the only bird I hit that week end. LLoyd did ok and my Dad shot his limit. But to hear him talk I was the only one that fired a shot.

We always stopped at a restaurant on the way home to get a bowl of chili. This was my Dads favorite traditions. The restaurant was always filled with hunters, truckers and local farmers. In the restaurant, my dad told everybody that would listen about the first pheasant that I shot. Of course I just sat there and smiled. It was wonderful.

I don't know how many hundreds of pheasant, quail and chuckers fell from my shot gun, but the first was always the best.

CHAPTER 2

A GIFT FROM GRANDPA

I sat in the large room looking at the ll other people in my group. The room has pictures of Viet Nam on the walls and I keep looking at them. Each picture looks very familiar and fills me with the old feelings about the war. All ll people were also in combat, even the therapist. It is our third time together and I am still amazed at how much that one year changed our lives. I am convinced that I was in as heavy Combat as the others but have handled the war better. I survived, went to college, was married for 17 years , have two children and have a good job. Some of the outward signs of the war is that I am divorced and usually change jobs after one year. But inside I don't seem to be as troubled as the others. I guess that is because I am tough.

Tough. The word makes me smile and remember another time.

I am nine years old and am very happy. It is summer and my brother and I are spending the summer on my grandparents farm with my 4 cousins. The reason I am happy is that Grandpa picked me to go with him for two days. He has property near the Canadian border were we will leave the steers that are in the back of the truck. Then we will spend two days fishing. Grandpa is about 60 and the toughest man I know.

The day is warm and the truck smells of hay. My door only opens from the outside and the drivers door opens every time we go around a corner. I'm sitting on a grain sack because the springs on the seat are exposed. We are flying down the road at about 30 miles an hour and I have to grab the gear shift every time we go up a hill to keep it in gear.

Grandpa is singing his favorite song. He sings, "Oh the bullfrog in the pond and the bulldog on the bank." Then he spits out the window. Every time he does this I laugh and that makes him smile. The second time he sings he reverses the bullfrog and the bulldog. This makes the song very funny.

"Steve, here is what we are going to do", says Grandpa. "We'll brand these steers then go to Sheep Crick and catch us some dinner". "What if we don't catch any fish, Grandpa?" I answered. "That is why I brought some food", He replies. I'm amazed.

When my parents go for a trip we pack the car with food and water and extra cloths and all kind of things. Grandpa had told me to put on my shoes, throw two blankets in the truck and he had made two sandwiches and boiled some eggs. I liked the way he packed, less bother. The sandwiches are mostly mayonnaise and the eggs are runny in the middle. That is grandpas favorite lunch. I'd had it many times and it was awful. I sure hoped we catch fish.

When we arrived at the ranch, Grandpa unloaded the steers and started a fire. He pulled out a branding iron from the old barn on the property. He said to me, "This old farm used to belong to your Uncle Jack. It a tough of a life up here for his family and he moved to the city. So, we use it". It was a wonderful place. The air smelled good and it was a beautiful day. All around the ranch were hills and mountains full of pine trees. The barn had one side covered with Deer horns. Uncle Jack had to hunt deer to feed his family. It was obvious that he was a good hunter.

"O K Steve, you hold the Steers while I brand them". This looked like it was going to be fun. Grandpa threw a rope around the first steer and pulled him near the fire. Then he pulled the feet out from under him and tied his legs together. "Now hold him so I can brand him". Grandpa said. When the iron touched the steer he started to kick. I wasn't going to let go and have grandpa bring one of the other kids next time. The steer kicked the crap out of me. I spent most of the time in the air. I didn't give up.

By the time the last steer was done, My cloths were torn and I was covered with dirt. Grandpa turned to me and smiled. I new he was impressed. He said, "you're pretty tough for being so skinny. If you want to make it in this world you got to be tough. Now lets go fishing."

We drove to Sheep Crick and Grandpa grabbed the two fishing poles from behind the seat of the truck. I didn't know they were there. Then we started fishing. I caught a bullhead and grandpa didn't catch anything. By the time it was dark we had walked miles and Grandpa was really mad. My father wore a suit to work and never swore. Grandpa wore dirty black pants, suspenders, a cotton long John shirt that used to be white and a red hunting hat that was dark around the sweatband. Grandpa swore.

Back at the ranch, Grandpa got out a can of coffee and put water in it. He put the can on the fire and boiled the coffee. I was amazed and asked why he didn't have a coffee pot. He said this was camp coffee and was better. We ate the awful sandwiches and the runny eggs and drank the coffee. Actually we chewed the coffee. It was full of grounds and very bitter. Then we laid out the blankets and slept in our clothes.

Grandpa immediately started snoring. I looked at the stars and listened to the coyotes yell. Sometimes I couldn't hear them because Grandpa snored so loud. At 9 I didn't know that coffee would keep me awake. Especially because that was the first coffee I had ever tasted. I sure hoped we would catch fish tomorrow.

As soon as the Sun came up, Grandpa awoke. I had just fallen to sleep. I had spent most of the night watching the moon crawl across the sky. I told grandpa that his snoring had kept me awake. He told me that he didn't snore. No matter what I said he would not admit that he snored. To this day I don't know if he was kidding or if he believed he didn't snore.

We got into the truck and began the drive to Sheep Crick. The road was dusty and covered the old red truck. The white dust made the truck look pink. Half way down the road, Two deer ran in front of the truck. Grandpa didn't even blink. It wasn't hunting season, they were doe's and he was going fishing. I thought they were beautiful. Grandpa told me that Uncle Jack used to shoot a buck and throw it up on this shoulders and carry it all the way home. He was a great hunter and he was tough.

At the bottom of the hill we turned right. We had turned left the night before. Grandpa had a plan. He decided that something looked funny to him the last time he was fishing up here and he wanted to check it out. I didn't know what we were going to do but I knew it would be an adventure.

We went to Sheep Crick and drove right by! Then we drove up a long hill and drove for about a mile and came to another creek. Grandpa called this a "crick" but I thought it was big enough to call a river. I asked him what creek this was and he said he was sure this was Sheep Crick. We turned around and went back to the lower creek. Then we got out and followed the Creek. We didn't take fishing poles. We walked for a short hike and then ran back to the truck. I was out of breath but Grandpa wasn't even winded. We jumped into the truck and drove like we were in a hurry. Grandpa's door was open more then it was closed.

Suddenly we stopped at the top of the hill and drove down a long dirt road. Grandpa jumped out of the car, grabbed his pole and ran into the trees. I was right behind him. The adventure was about to begin.

We came to a cliff and Grandpa said, "I knew it" and over the bank he went. I stopped. I am a little afraid of heights and this was a high cliff. There were just enough trees growing on it so I couldn't see the bottom. If I didn't follow Grandpa...I didn't want to think about it. Over the bank I went.

Grandpa did not slow down once. We were grabbing roots, trees, bushes and rocks to keep from becoming airborne. My legs hurt and my hands were bleeding but I was on an adventure with my grandpa and I had to be tough.

At the bottom was a great meadow. The grass was taller then me. Grandpa didn't even slow down. Into the tall grass we went. Grandpa yelled to keep my pole behind me so it didn't get tangled in the grass. I followed the path he made. There was no way I could keep up. He was 60 but he was fast!

I could hear a waterfall getting closer. By the time I got to the waterfall, Grandpa already had a worm on his hook and was casting. He told me to be quiet and not scare the fish. Just as I got a worm on the hook, Grandpa pulled in his first rainbow trout. I was amazed. Grandpa smiled and said, " The best fishing is at the bottom of a water fall. This is what I was looking for. Now get fishing before I catch all of them."

Grandpa caught six fish and I caught four. Grandpa was very pleased and made a big deal about me catching the biggest fish.

He yelled at me for not knowing how to clean a fish and said I was a city slicker. After he showed me how to clean one he gave me his knife and had me clean all the rest. I had not held a knife that was So sharp. The water was cold and the fish were slippery but all I could think about was that I didn't want to cut myself. If I did I was afraid I would cry and so far I was doing pretty good. Grandpa still thought I was tough, even if I was a tough city slicker. You couldn't be tough if you cried.

Grandpa cut a branch from a tree and put the fish on the branch. Then he gave me the branch and started up the cliff. I held my fishing pole in one hand and our dinner in the other. Climbing that hill would take both hands. I didn't want to break the fishing pole and I didn't want to get the fish dirty. I put the pole and the fish in my left hand and followed grandpa. By now he was almost to the top.

By the time I got to the top I was ready to drop. My left hand ached, my legs ached and I thought I would get heat stroke. Grandpa was leaning against the truck smoking one of his home made cigarettes. He could roll it with just one hand. Incredible! He looked at the fish and smiled. Then he took off his hat and put it on my head. I almost laughed because he always wore his hat and his head was white. Grandpa was bald and the white skin made him look very bald. Grandpa said, "You'll do all right, your tough! Let's get into the truck". And down the road we went. Back at the ranch, we parked the truck and Grandpa cooked four of the fish. In my life, before or after, I have never tasted food as good as that. Grandpa recooked the coffee but I didn't want any. I did drink water.

After dinner Grandpa wanted to walk the ranch. He wanted to look at the fences. As we walked he told me about his life growing up and what his father was like. I didn't think I would have liked his Dad. He also told me more about Uncle Jack and how he swam across the Columbia River just to win a bet. Of course, I knew Uncle Jack but to me he was this barrel chested, soft spoken man that usually smelled like sweat. Grandpa liked to talk to me and I liked to ask him questions. This was new for me. At home grandpa didn't seem to have the time to talk to kids. I was having a wonderful time. When he walked he looked like he wasn't going very fast but I always had to hurry to keep up.

After walking for hours, we got back to the truck. Grandpa cooked the rest of the fish and sat back to relax. I don't remember anything else. I was asleep.

I was in Viet Nam for one year, one month and one day. I saw close friends die and I killed people. I was 20 years old. Twenty two years later I am still having flash backs and am still hurting inside. But, then and now, I can feel those feelings and I can still function.

When the memories of the war come crashing into my life, there is also a voice. An older memory and an older voice that says," You are going to survive this pain. You are tough!"

CHAPTER 3

DEATH FEAR

Monsoon season, 1968, Vietnam. We were completing one of our typical days maneuvers. These represented going somewhere, driving around for awhile and returning to someplace to spend the night. I usually looked forward to these trips because it was more interesting then sitting and waiting. Most of my early days in Vietnam were very boring. Later, I realized the leaders were sending my platoon of five armored personal carriers to areas that they thought were occupied by the enemy. We would parade around like the red coats in the revolutionary war waiting for the enemy to attack.

Our company consisted of three platoons of armored personal carriers (tracks). Each platoon had 5 tracks and were occupied by four or five solders. Each platoon had a given area to guard. Our platoon was first platoon and we typically spent our time in the rice paddies. The second platoon spent most of their time in the city of Tuy Hoa and the third platoon stayed at our base camp. Their job was to guard the command post and an artillery group.

On this day, we again did not see the enemy. Our return trip was next to Chop Chi Mountain and through Chop Chi Village. I had not been in this part of my area in the central highlands.

The monsoon season is wet. The mud in the paddies is thick, smelly and deep. The tracks swim in the mud. I was a driver and very proud of my job. To be a driver took skill and intelligence. I was 20 years old.

As we approached the village, one of the new drivers drove through a part of the rice paddy that was too deep and he became stuck. A second track pulled over to help him and also became stuck. We chained the two tracks together and then ran a long chain to the other three tracks that were on relatively dry ground. The long chain was stored on my track. I carried it to the stuck vehicle. As I got close to the track the mud was over my waste. I remember standing in the rice paddy holding the heavy chain with the smelly mud at my waste and looking around and being amazed. The rice had just started to grow. It was about 3 inches tall and I found myself waste high in a sea of beautiful light yellow green.

Eventually, we pulled everybody out and were about to get on our way. The procedure attracted a lot of local attention but especially one older couple. The woman had been yelling instructions to us. Of course, even if we could have understood her language we would have ignored her. As we began to drive away she sat on the rice paddy dike and cried. She was very upset and kept pointing behind us. Then I looked back. The beautiful sea of green had turned into a plowed field. The entire crop, probably her families future income, had been destroyed. The farm boy in me felt sad. The solder said this is war and went on.

Two months later I was in this same area. This time it was night in a battle with the North Vietnamese Army. I spotted two flashes in the Chop Chi village and we began to receive rocket rounds. One round was fired from the same dike where the lady had stood and cried. My best friend died that night from a round from a rocket launcher.

After that battle, our leaders determined that the platoons should rotate duty. We would spend two weeks in the city, two weeks at the command post and two weeks in the rice paddies.

In the entire year I was in Vietnam, the third platoon did not see any combat. The second platoon had the misfortune of hitting a few land mines but the first platoon was always fighting. We talked about it often when we would get together as a company. The third platoon believed it was because they smoked so much pot that Charlie thought of them as one of their own. My platoon was in so much combat that most of us were afraid to take drugs.

It was amazing. Each time we went to the rice paddies the combat would start. As soon as we went to the other locations, it would stop. I remember warning a friend in another platoon that Charlie was all around us on our last mission and he laughed. He said that Charlie would wait for me and leave him alone. He turned out to be right.

One day we were on a mission as a company. This was rare. As we returned, we had to cross a dike that was about 15 feet high. I was point driver for our platoon. The Second platoon was ahead of us. Point driver was a very prestigious position. Only the best driver with the best track commander was allowed to be point. I began to drive over the dike where the climb was toughest. A rule of thumb is to take the hardest, most difficult path between two points. The easiest path could be mined. The second platoon began take the easy approach.

Suddenly, the point track for the second platoon exploded and lifted into the air. It went about three stories high and came down on its nose. Armored personal carriers when loaded weigh over fifteen tons. This had to be a Russian made mine. People were thrown everywhere. I immediately turned my track toward the heavy brush and the hillside next to the rice paddy. As the track turned I heard the 50 caliber machine gun next to my head slip a round into the chamber. The track commander was thinking the same as I was. Charlie was near by.

After we decided we were not going to get hit immediately, he and the two gunners climbed off the track to help with the wounded. I was left alone to sit on the machine gun and to listen to the radio.

I yelled that I would control the dust off and that they should bring the wounded to me. I then got on the radio and talked to the dust off chopper. He was about 10 minutes away.

Four men carrying a wounded solder on a stretcher ran toward me. When they were about 15 feet away there was a terrible explosion. The left front solder had stepped on another mine. Mud and dirt splattered my skin. The man that stepped on the mine no longer looked like a person. His legs and arms were gone. He was dead. The man on the stretcher lifted his arm and he died. A third man was in very bad condition. Most of his face, shoulder and one arm were gone.

I called in the chopper and the wounded and dead were taken away. The third man died on the ride to the hospital.

That day I mentally began to prepare for my death. I wrote to everyone and said goodbye. Even the earth could kill you and I had four months left in country. I knew I wouldn't make it. I became very scared. I didn't want to die.

About a month later we were in town. The next day we were scheduled to go into the rice paddies again. Everything had been fairly calm for awhile. I was still lead driver. The drivers listen to the radio as well as the track commander. I was in track one-seven. The one represented the platoon number. There were 5 tracks. Of course we didn't use one-one and one-two had been blown up so many times that it was called one-seven. It was supposed to be lucky.

As we set up our parameter an oriental voice came over the radio, "One-four, one-seven, you die tonight". My heart stopped and my blood froze. I was so scared I could hardly speak.

I didn't sleep but stayed up all night, sitting with each person as they stood their guard. The next day, I was still too shook up to relax. The second night I again stayed up. The second day I slept all day.

It took about two weeks for the enemy to attack. When he hit it was horrible. He surrounded us and attacked with everything. Two men died that night and almost everyone was injured. I earned my purple heart that night. BUT I didn't die!

The few days later we found pieces of paper spread around the country side. On one side was Vietnamese writing. On the other side was English. It told of the glory of the North Vietnamese Army as they defeated the 173rd Airborne Company. They told of many armored personal carriers being blown up and many Americans killed. One word was used over and over, revenge. I think they wanted revenge for killing so many of their people, for being in their country and ruining their fields.

Finally, after months of small battles, I got to go home. I came home a changed person. Not so much from the memories of the war but from my outlook on life. I can not shake the fear of dying. It creeps into my relationships and my day to day activities.

About 2 years ago I received a call from my mother. She said my Uncle Burt had cancer and that the doctors did not expect him to live very long. I didn't know Uncle Burt as well as my other uncles. I knew him as a warm good person that had great stories at all the family reunions and a wonderful laugh.

My mother told me he was going to come for a visit soon.

When he did come, he stayed at my Uncle Jims. I drove to see him. I expected a person that was in pain and would be trying to hide his fear of dying. I expected him to give us all words of wisdom and tell each of us goodbye. I didn't really want to see him.

When I arrived at uncle Jim's, everyone was downstairs watching a football game. When I walked into the room, Uncle Burt got up and shook my hand. It was a strong solid handshake. Then He smiled and I felt that strange feeling he always gave me. A feeling that life was good and that he was especially glad to see me.

We all sat around and laughed and enjoyed each other. I ate too much and completely enjoyed being with my relatives.

Uncle Burt was not sad nor scared. He was not especially brave either. He was just himself. As I drove away, I was amazed at everything that had happened. I realized that Uncle Burt was going through his final days the same way he lived his life. He was self confident, and generally a happy person. He lived his life with pride and self control.

He died last year.

Uncle Burt, you left me with something special. I do not have to let death stop me from enjoying life. In fact, even when faced with death, I do not have to let that rule my life.

I also know another thing. If there is a hell, it is like being in Vietnam waiting for death. And if there is a heaven, it will be sitting around a fire with all my relatives listening to you tell funny stories.

CHAPTER 4

INNOCENCE

A few weeks ago I was traveling by car to Omak to present a proposal for a possible project. I was driving alone and it was very hot for a June day. I have always loved to drive in Eastern Washington. The sky was blue and the air was clear.

When I arrived, I raced into a local Burger King and put on my suit. I drove to Omak Hospital and walked in the front door. I had an appointment with the administrator but did not nail down a time. I was not sure when I would arrive at Omak. "I'm sorry, Mr Mack will be in meetings all day", said his tall attractive secretary. "Please let me have your brochure and I'll make sure he sees it." I had just driven 5 hours. I gave her my brochure and my best professional smile and walked out of the hospital. Then I drove back to Burger King and changed into my tank top and jeans. I ordered a burger and a milk shake.

I decided to drive to Winthrup. To get there I drove over Loop Loop Pass. It is a beautiful drive. At the beginning of the pass I saw a sign that indicated there was a historical marker ahead. When I arrived, I read" The great sheep slaughter. Immediately my mind raced back 15 years.

Fifteen years ago. Was it really that long ago? I was going on vacation at Curlue lake. My vacation started a week before my wife's vacation so I took the kids by myself. We had taken the North Cascades Highway and we were having a great time. The rule was that we would stop at anything that interested us. We had just spent about an hour in Winthrup. Brian was in the front seat and Heather was in the back. Heather said," Dad, I really like driving with you. This is lots of fun. I smiled and looked in the rear view mirror. She was a beautiful blond girl that had an instant tan as soon as summer arrived. She had a bag of potato chips on her left and a bag of cheetoes on the right. On her left knee was a box of red licorice and on her right knee was a box of black licorice. In her lap was a coke and on the floor was a six pack (minus one) of coke. On her face was a big smile. Mom would never let her get away with having her own goodies.

Brian was in the front seat. All the windows were down and his blond hair was flying in the warm breeze. He hadn't started school yet and was full of ambition and confidence. Now he is 20 and six feet five inches tall. So many people have told him that he "can't" that now he says to himself," I can't". At that young age he could do anything. He hasn't yet reached that age where he has found himself. To me, that means that he finds out that he can handle the world and all the problems and bills.

Then he finds that little boy in himself and begins to once again enjoy life.

"Dad," Brian said, "Are there fish in the lake where we are going?" "I think so. The last time I was there was with my Grandpa when I was Heathers age. We didn't catch any fish that day". "Tell me more stories about your grandpa" Brian answered. Before I can talk he proudly explained that he will catch the biggest fish. "Maybe even a rainbow trout!" Heather answers that she will catch the biggest trout! "You will not". "I will." "Will not". "You can't because you are a girl and only boys can catch the biggest trout". Heather put out her chin and answered, " I will too".

Ahead was a sign. I read it to the kids. "Historical marker ahead". "Oh good", said Heather. We stopped at all the historical markers. When we arrived, we jumped out of the car. The air smelled of pine trees and hay. The sun was hot and there was a breeze. The kids got on each side of the sign and I took a picture. Then Heather began to read the sign. "The great sheep kill."

Fifteen years later I smiled at the memory of my beautiful children. They grow up and they change. At the time everything is so busy. Then one day you miss your children. Suddenly there are two young adults standing before you. The children are gone, forever. If you look closely you can see the children in the adults, but you will never find those wonderful little friends that meant so much to you.

As is my typical pattern when I am alone for a long time, I begin to think about the war. I don't think about it on purpose. It just happens. Remembering my children, my mind goes back 22 years to other children. I stop and think about my feelings. I feel angry at authority, deeply sad and guilty all at once. Why?

We spent most of our days sitting. At night we did most of our fighting. During the day I would read or write letters or I would talk to the Vietnamese children. We did not frighten the children as much as the adults and they were always around. I guess when they were with us they were safe and we gave them food. My favorites were a couple of brothers. The oldest was about 11 and the youngest about 9. The older brother could speak English. His younger brother would whisper to him in Vietnamese and he would talk to us. The 9 year old was actually smarter then his older brother but was quite shy. They would run to get things for me and they would talk about their lives. Slowly the younger brother talked to me. His English was poor but I could understand him. He was funny and would laugh very hard when he made a joke.

About that time the leaders of our army came up with a great idea. They put a bounty on explosives that were laying around the country side unexploded. Many artillery rounds didn't go off and some mortar rounds. The Viet Cong would collect these and make land mines out of them. Some brilliant leader thought this was a great idea to stop the mines. Of course this was also during the time when the bombing was halted. When the bombing stopped, their supplies poured into the south. The Viet Cong were using Russian made mines because they were more reliable.

The day after this was announced, my two young friends walked into our area carrying an artillery round on each shoulder. I saw them and ran. These are very explosive and could have killed all of us. The boys walked into the area and flipped the rounds off their shoulders. I screamed as the rounds hit the ground. None of them exploded. Our fearless lieutenant walked up to the boys and gave each of them a few dollars. He had put on his bullet proof vest and his helmet.

The boys ran to me to show me their money. They were "rich". Each day I asked them not to bring in the rounds. I began to give them more money for running errands. They still brought in the rounds.

One day only the older boy came in. "Where is your brother?" I asked. "He ka ka dow", he answered. This is a Vietnamese word that was used when we shot a Vietcong. It may have been their way of saying K.I.A. "What do you mean", I demanded. I was sure I knew what he meant. "He die", he said. I grabbed the boy by the shoulders. "Where. lets get a medic. We can get him to a hospital. Hurry, before its too late." He looked at me. No emotion showed on his face. He said, "too late. He gone."

The boy had dropped a round on its nose. There was hardly enough to bury. Yes, I still feel angry at our leaders who were so smart.

As a friend told me, "They were saving American lives. So what if a few gook children died." We were called "baby killers" when we got home. I guess we were. But we were not the only baby killers in the war. The Vietnamese also killed children.

I was walking around the town with a couple of my friends when a child suddenly looked at me and took off running. The town had narrow streets with card board houses on each side. The roofs were made of corrugated sheet metal. Entire families lived in these little houses. We liked card board houses because bullets went through the walls.

The boy that had ran away came back with a young woman. She looked at me and tears started running down her face. She said over and over," You came back, you came back". She took me by the hand and led me down the narrow streets. My friends thought this was funny and were making jokes about what a great lover I must be. The women couldn't wait to get me to their hooches. I yelled over my shoulder to hurry up. I didn't know where she was leading me. Besides, she was not a prostitute.

I stopped her and asked what was going on. She told me in very broken English that she was so happy that I was back and that she had a great surprise for me. I told her that I had just arrived in Vietnam and I had never seen her before. She went on and on about how happy she was. Finally I looked her right in the eyes, face to face and said I didn't know her. She stopped and looked at me from the side and from the front. She held my face very close to her. Tears began to well up in her eyes.

Finally she stood back and said" It is not you!" Quietly she walked to her house and I followed. When I got there I sent the boy to get a coke. I sat in the shade and she told me that she had fallen in love with an American that looked just like me. He had taken care of her and bought her many gifts. Her family had been angry at her but she loved him. He had told her many times that he was going to take her out of the war and to America. When his tour of duty ended he told her he would return to get her. After he left , she found out she was pregnant.

I was sorry that I was not him but the fact was, I wasn't. She brought out her boy. He looked huge in her arms. She was less then 5 feet tall. He looked very Vietnamese except his skin was pale. He did not look like me. He was only a few months old.

About four months later I was in that same area and I walked to that part of town. I went to her house but she wasn't there. The boy was there and he took me to her new home. When I found her, she had lipstick on and was working as a prostitute. I was surprised and asked her what happened. She said that was the only way she could make money. I asked her how her baby was doing. She looked at me and said that she didn't have a baby. As I walked away I asked the boy what had happened. He said that she had "much shame". That her family forced her to be a prostitute and they had killed the baby.

Somehow, I felt that it was my fault. I blamed myself. I'm not sure what I could have done but I felt as guilty as if I had been the father. I was amazed these people had children. It seemed to set them up for pain. Why have children that would die?

Vietnam aged me in a way that made me old inside and outside.

The most dramatic was after an all night battle. I sat on the back of my track and held the enemy machine gun that we had just captured. My friend used my camera and took my picture. I was shocked when the picture came back. I know what I will look like at seventy. The picture was of an old man. We had fought all night and I had just loaded three of my friends on a chopper. They were dead. I sat there struggling with control over the pain. I could not morn. I could not think about how sad it was to send three friends home dead. I could not think about the four men on the previous chopper. all wounded. One with his legs gone from the knees down. One with a piece of metal in his head. All bleeding badly. No, I could not think about that. I had to go on. I had to survive. If I dwelled on them, if I let my guard down, I would not be ready when the enemy came back. I began to feel numb. Numb was good.

My friend pointed at a lump on the ground. "Look, its old Granny". Every time we had set up our night guard at this position, we had been visited by the family in the nearby grass hooshes. These were farmers. They had many children. I think there was one at every age from birth to about 16. Their grandmother lived with them. She was ancient. She did not have teeth and she always begged for food. We called her "old Granny".

She was laying on the ground, dead. Her family was stepping over her body. No one seemed to notice her or do any thing about her.

Suddenly the mother walked around the corner. She was crying and carrying the four year old girl in her arms. The front of her clothes were covered with the girls blood. The girl was dead.

She stepped over Granny and walked up to me. She said something in Vietnamese. I answered, "We did not kill her, She was killed in the battle, by the Viet Cong." She answered in Vietnamese. I could not understand her words but I understood her meaning. I yelled," We could barely keep ourselves alive. We couldn't protect her. We did our best." She said something else through her tears and I yelled that they were all stupid. Why did they have children. Didn't they know they were living in a war.

Didn't they know they would lose some of them? It is not my fault. They shouldn't have children if they couldn't handle the pain.

In a battle between two elephants, some of the grass gets trampled. Some of the innocent grass. It was war. It was the way it was. But that does not take away the pain.

Loosing a friend is very very painful. Seeing people killed for no reason is very sad. But seeing a child die in unbearable.

CHAPTER 5

DREAMS

In my memory, Vietnam is very close to me. The memories of Vietnam are very fresh and clear and seem to be more recent then memories that happened after I came home. Part of this is caused by dreams. Dreams related to Vietnam keep the feelings and the emotions of war close to me.

Of all the multitude of nightmares that have destroyed my nights, there are a few that stand out and are repeated often. Following are a few.

My first nightmare happened when I was in Vietnam, close to the end of my tour. I am at the bottom of a river, holding on to a mossy rock. The rock is my life. If I let go of the slippery rock, I will die. I am able to bring my head up to get air but if I stay up, the rock starts to slip from my hands. Holding on to life is very, very difficult. Suddenly a foot with Vietnamese sandals appears in the water. The foot comes down on my hand and grinds on my skin. I can't defend myself or I will let go of the rock. I can't even scream. As my hands come loose from the rock, I wake up. I always awake gasping for air.

Not all dreams are about the war but have the same feelings I had when someone died.

I am standing on the sidewalk of a dead end road. At the end of the road is a brick building. This is to my left. I look across the street and see my son. He is about 7 or 8. I scream at him to be careful crossing the street.

When he is in the middle of the street, I see a truck coming down the road. I scream but my son does not hear me. The truck hits him and he is thrown into the air and down the road. He passes over my head about 10 feet high. I am helpless to do anything. As he passes me, I can see that he is awake and seems to be OK. He holds out his arms to me and yells, "DAD".

I feel relief that he is not hurt. Then I realize that he will hit the building. The scream catches in my throat as he hits the brick wall. His head smashes into the wall and blood is thrown everywhere. I always awake from this dream screaming my sons name. I never go back to sleep.

Vietnam is a very beautiful country. In one dream, I go back to visit as a tourist. I go to the places where the old battles were. I am telling someone, usually a woman, what it was like. Suddenly, my best friend walks out of the grass houses. I can't believe it. I run to him and say, "Joe, You are alive. I can't believe it. I am very, very excited. Joe, why are you still here?" He smiles and says that he likes it here. I tell him that everyone thought he was dead. I just want to sit and "rap" with him the way we used to. I am so relieved. He hands me a warm beer, just like we used to drink together. I have tears in my eyes, I am so happy to see him.

I hear a rifle fired and I turn to see a Vietcong shooting from the bushes. I feel confused and unprotected. I am scared. I turn back to Joe and he is hit in the chest. NO! I scream. No not again. You are alive. No do not die! I always awake from this dream sobbing.

Joey, My "Significant others" son and I are very close. He is a wonderful boy. He is 10 years old. He likes to call me his stepdad. Once he asked me to change my last name so we would have the same name. In my most recent dream, he and I are in a strange country. The air is very dry and it is hot. We are walking up a hill and suddenly a group of people appear. They have one goal, that I immediately realize. They want to kill Joey. They can't kill me. I am protected for some reason. But they can hurt me.

They start chasing us and I grab Joey and put him on my shoulders. I am very scared. There are too many of them. They are wearing hoods and I can't see their faces. They are all carrying clubs. I imagine them cornering Joey and smashing him with the clubs. I turn to see how close they are and my ankle twists. Both Joey and I fall. I can not worry if he is hurt in the fall. I roll and turn in time for the first person to get within a few feet of me. I dive at him and take his club. I bring it down on his head just as the others arrive. I put myself between Joey and the mob. I am hurting and even killing them but there is so many. Blood is flying all over and the ground is red.

When I am hit, I don't feel it. I am afraid they will break my arms and I wont be able to fight. I look back to see Joey fighting with one of them. He seems to be winning. I back up to help him, as a club is brought down on my shoulder. It hurts and I think something is broken. I can not give up.

I move closer to him. I see someone coming up behind Joey with a club raised over his head. I wake up screaming, No!!!!

I will not go to bed without checking to see if all the doors are locked. I have to do this because of my dream. I am asleep in my dream. I hear the sliding door open and I sit up in bed. I hear movement and I slip out of bed. In the dark, I see a figure. He is a Vietcong and he is hunting for me. I back up to the closet where my shotgun is kept. I do not keep bullets in the house. But, in the dream, I have bullets. I can not put them in the gun because it will make too much noise. I sneak back to the living room and I see the enemy moving behind the couch. I back down the hall. Suddenly he yells, "Steve, where are you. I have come to take your life. When he starts talking, I load the shot gun. " I have come to take revenge. You killed my father." I yell back that he is as stupid as his father and he will be the one that will die. He turns and fires his automatic rifle. The bullets easily go through the walls.

I crawl on my stomach toward my enemy. I can hear him breathing. I swing my body around the corner and face him. I fire my weapon.

Panic. I do not hit him with a shot gun blast but a BB hits him in the chest. At first he winces because he thinks I have killed him. Then he looks at me as if to say, "what is this?" and then he smiles. He brings the rifle level with my face. I throw my body backwards violently to escape. I usually find myself on the ground next to the bed when I awake from this dream. I am in a cold sweat.

These and others like these, do not stop. I used to wake my wife up to tell her the dreams. I hoped that telling the dream would make it go away. I've told therapists and they are very helpful telling me what the dream means. Usually, I am reliving an emotion or a feeling that I felt in the war. This is great information but no one can tell me how to make them stop.

Time is supposed to help. It has been 22 years. 22 years of dreams.

I guess it could be worse. I could be the poor person that sleeps next to me.

CHAPTER 6

THERAPY

It was fall, 1989. A typical Seattle fall day, raining. I was going to my group therapy session at the Vietnam veterans outreach program.

I hate therapy and group therapy. 20 years ago I was in Vietnam. I spent one year of my life there. One year out of 42. I keep telling myself that one 42nd of my life can not govern so much of what is happening to me. Imagine one 42nd of a cup of sugar! It could hardly be seen. How could this be influencing my life?

I hate therapy. I also dislike therapists. I have seen 4 and they all talk the same. They all talk about "stuffing it" and love to use words like "disorder, Working on issues" and post traumatic. They also don't like to answer a question without asking a question. That usually pisses me off. But I have to go. Why? Because I know something is wrong. Group therapy is supposed to be the best thing for me. The problem is that VA therapy is free to all veterans. This takes away the desire to work on problems. The therapist can not help. They only make you look at things that you weren't seeing. Then YOU have to go back and work on a solution. Working on solutions is hard and takes energy. When it is free, why work? I will come back next week and try to deal with this feeling later. Thus, group therapy doesn't always work and it takes forever. This group lasted for one year and only broke up because the therapist quit and went to work somewhere else.

I walked into the group and sat down. The theme of the group is to look at todays Issues. If we are having problems, then we talk about it. The therapist gets real excited if Vietnam is brought into the discussion.

The start of the session is called "checking in". This is a time when each person has a chance to tell what happened last week. Most of the sessions last 2 hours and checking in takes the entire 2 hours.

It was my turn. I answered that the last week was bad. I had spent most of the week trying to start fights with my "Significant Other". My goal seemed to be to break up with her. During the argument I always knew I was right but later I would realize I distorted the problems.

The therapist asked what the first argument was about? It started on Friday. She sells real estate and had been showing property. A couple she shows houses to had decided they wanted to build the house of their dreams. This evening she was only showing to the husband. They were looking at property where he could raise horses. Because of all the rain they had taken his four wheel drive truck. She was glad they had taken his truck because they had almost gotten stuck. She had arrived home well after dark. I asked her if she had gone to the office to write up an offer. No they had gone from the last piece of land to home.

I became very agitated and asked if she had been out after dark. She said they were looking at as many sites as possible and, yes it had gotten dark before they left the last lot. Then I asked her if she had been carrying the mace I got her. She said she always left it in her car and no she did not have it with her.

I became very upset and told her that she could have been raped or murdered. She said that she was a good judge of character and that she was not afraid. Her proof was the experience that night. She had not had any problems.

My therapist then asked, " How did you feel when she said that?" I replied I felt like I wanted out of this relationship. That I couldn't handle it if she were raped.

One of the other people in the group said, "what's the big deal? It's just sex and people recover from that every day."

I came up out of my chair and wanted to kill him. I screamed that it wasn't sex. It was violence and it was horrible. That people with his attitude should be force to watch a rape. Then he would grow up to the facts. I was very angry.

The therapist then asked if I had seen a rape. I answered yes, in Vietnam. That I hadn't thought about it for years and I really didn't want to discuss it. What I was trying to do was find out why I was trying to end my relationship. He said that he wanted to explore this area further. I was pissed. Of course he wanted to explore this further. It was right up his alley. Vietnam related issues!

"It was 20 years ago", I said. It was during the last few months of my time in Vietnam and we were guarding a bridge. We did that for months. All day guard a bridge. At night run around chasing the enemy until he caught us.

We did not have race problems. My best friend was George Alvin Tyler the third. He was black. We didn't see color. Just who would stand beside you when there was a fire fight. There were two friends, one was black and the other was white. I don't remember their names but I do know they were named backwards. The white guy had a name like Leroy Jefferson and the black guy was named something like Mike McDonald. The white man was over 6 feet tall and weighed at least 200 pounds. The black man was taller and about the same weight. They liked to go to the rear area and start fights.

One day a young girl carrying two baskets of rice on a pole over her shoulder came to the bridge.

We only saw two types of Vietnamese women, the prostitutes and the civilians. This was not a prostitute. Leroy stepped in front of her and asked if she was looking for a good time. She couldn't speak English. She put down her load and handed her papers to him. He looked at the papers and handed them back. He said, "here, you stuck up bitch". I was on the other side of the bridge and was watching. the black man walked up behind her and took her baskets away from her. She started yelling loudly in Vietnamese. He looked at her for a minute and hit her in the face with his fist. It knocked her down. As she started to get up, the white man knocked her down. He kept pushing her toward the side of the bridge. The black man grabbed her top and ripped it open. She was yelling and kicking. I stood on the other side of the bridge and I watched.

The white man held her shoulders and the black man rapped her. She was screaming and trying to get loose. The black man hit her hard in the face with his fist to make her shut up. He did it over and over. Then they traded places.

The black man started laughing at his friend. He did not have an erection. The white man started hitting the girl. This aroused him and he raped her also.

I stood on the other side of the bridge and I watched. I didn't walk away and I didn't try to stop them.

When they were through, she picked up her cloths with out saying a thing. She put her blouse on and stood very straight. Her face was swollen and red. Her nose was bleeding. She said something in Vietnamese and both men laughed. She picked up her load and started to walk. Her legs were shaking. She walked by me. She looked me in the eye and she spit on me. I turned and walked away.

About 8 of us had seen the rape. No one talked for the rest of the day. That evening before we started our first guard, the sergeant said that he was in charge and should have done something. He said that if he had, those two would be in jail. "It is against the law to rape anybody, even gooks".

My therapist asked what I did next. I told him I forgot about it. It went away. He asked why I hadn't stopped the rape. One of the other men in the group answered that it was obvious. That I couldn't stop it. I had to live with those men for the remainder of my tour. I kept them alive and they kept me alive. In that situation you couldn't afford to make an enemy. Not if you wanted to go home. It was life and death.

The therapist asked how I felt about that answer. I said that he was right, but that didn't take away the horror of what happened.

Then I turned to the man in the group that had made me so angry. Every day prostitutes crossed our bridge. There was plenty of sex and it cost five dollars. Those men did not want sex. They wanted to hurt someone.

The therapist then asked what I would say to those men if they were here. I said that I would not look at them nor talk to them. That was what I did for the rest of my time with them. Would I be angry? I will be angry at them for as long as I live.

Then the therapist asked what I would say to the girl if she were here. I said that is easy. I would ask what she did to get revenge for the rape. I would ask how many Americans she killed before the war was over. "Did you suspect that she was a Vietcong?" "No, I doubt if she was. But, I would have expected her to be after that day."

Then the therapist asked me to talk this over with my Significant Other. Tell her how much it scares me to have her take chances. Tell her about the horror of rape. She told her about the session. She said I went home.

The next night I told her about the theropy session, and she said that I shouldn't worry. She was a good judge of character. I would be the one raped. And I would just handle it. I think this is a great difference between men and women. My theory is that women go through such pain when giving birth that they have a built in "something" that lets them handle things. If they didn't, women would only have one child. Men do not have this capacity. When I talk to men about the traumas of Vietnam, they are usually understanding and will even try to help. All women I have known always say, "Just handle it and go on with your life."

I told her it was very important to me and to our relationship that she please be careful.

She said, "ok" and got up to go to her next appointment. I grumbled about another late evening appointment. She gave me a kiss and was gone.

I looked over at her side of the kitchen table where she at just sat. There, on the table sat her mace.

CHAPTER 7

LEADERSHIP

In my life I had never been so tired. Tired to the bone. Each day I would reach into a new level of reserves to keep my tired body going. I was in the second week of sleeping no more then two hours every other day. I was starting to hallucinate. Actually, when you get that tired you begin to dream. You have to dream. So you do it when you are awake. After awhile, reality seems to become a dream.

I was in my fifth year of College and I was working on my masters degree. In Architecture, the major part of the fifth and last year is working on a Thesis project. My project was a combined church and a school.

Looking back on those years, I realize that I was "popular". At the time I didn't think so. I was married and had two young children. Our income consisted of tuition money given to me by my father-in-law and from the GI bill. The GI bill paid $316.00 per month.

I was the president of the student chapter of the American Institute of Architects. Our class had 60 students. Sixty future architects.

My professor and I did not get along. He was a poor leader and had a secret agenda. In my mind he wanted to be the dean and would walk on anyone to get there. He was not a good instructor. I tried to get along because he was the key to graduation. But something got in the way. He made me very angry.

Two weeks before my thesis was due, he stopped by. He laid his cup of coffee on my desk. He always did this when he had something profound to say. My design was almost complete. I was ahead of schedule and had the model almost built. He disagreed with my basic premise that the school and the church should be separated. He wanted to move the church into the building. He explained this and I argued. He grabbed the roof of the model and ripped it off. I couldn't believe it. I was using money to buy supplies that should have been food for my children.

He then ripped the model apart. As he did so, he talked quietly about how the design should look. Finally, I asked him if this was his design or mine. He looked very angry but continued to talk softly. He took my finished drawings and sketched over them with an ink pen. They were now ruined. Then he turned and stated that if I did not change the design the way he wanted and, if I did not get completed in time, I would flunk the class. He looked at me and said that I was good at talking to people and making friends.

In as calm a voice as I could use, I told him that the drawings and the model would be completed and that he would be forced to pass me. More then any urge in my life, I wanted to kill him. He was disliked by the other students as well. About 6 of my fellow students had witnessed what had happened. They rushed to me as I turned and looked at my model. A powerful feeling of despair rolled over me. One friend said that he would help me. Then the others said the same.

Leadership 39

One of my fellow students had brought in a lump of clay from an art class. He roughly modeled a face of the professor. During the days to come, every student that would pass the lump of clay would touch up the bust of the professor. After a few days, the owner of the clay dug out the insides and added a handle. He then took it to the art department and had it fired. A perfect cup with the Professors face came back.

I began to rally the other students around me. "We have got to do something about the Professor. We can not let him get away with this. We must do something." All agreed. When you put the energy of sixty creative architecture students together, nothing can stop you.

One of the students had a hobby of making motion pictures. We made a film. One student put the cup on his head and wrapped a coat over his face. He looked like the professor. In the film, the fake professor arrived at my desk and spilled his famous cup of coffee all over my drawings. He then proceeded to destroy my project. I tried to stop him and he knocked me down. In the end I was left with my head on my desk with a destroyed model. The Professors last words were, "You will never make an Architect."

We took the film to the Dean of Architecture. We showed him the film and explained what was happening in the fifth year design lab.

Most of the class went to the meeting. On the final days of design lab, I was having trouble knowing what was real. Everything seemed to swim in a fluid. I would watch the sun go up and the next thing I knew I watched it come up again. The other students would finish their projects and come over to my desk and help me. One student stayed up all night helping me. One half hour before it was due, the entire class was working on my project. Sixty students. At two minutes before it was due, I proudly took the project into the professors office. He looked at me and said, You did it, too bad I was looking forward to flunking you. When I got back, the student that sat behind me, a very talented designer, shook my hand. His girl friend threw her arms around me. I walked to the telephone and tried to call my wife. I could not make my memory work. I could not remember my phone number.

I started walking the two miles home. I just walked. I watched the rocks pass under my feet. I do not know how long it took to get home. I walked into the house and went straight to bed. I woke up 24 hours later and ate. I went back to sleep and slept another 12 hours. The remainder of the summer I needed at least 12 hours sleep a night. My dreams were full of killing. I dreamed over and over that I killed the professor. When my report card came I carefully opened it. Next to architecture design class was a "B". On graduation day we were all dressed in our robes getting ready to graduate. One of my friends ran up to me to tell me the great news. The Professor had been asked to resign.

I can not escape my strong feelings about poor leaders with hidden agendas. My feelings are powerful. To other people, these feelings are too strong. To me they are life and death. I have quit many jobs because I could not deal with the incompetence of the leaders. I try to get rid of them. If that does not work, I leave. Poor leaders make men die.

June 6, 1969. Vietnam. Our mission was to meet with a patrol of ARVNs. These are the Vietnamese soldiers. We were late and it was getting dark. As we tried to climb over a rice paddy dike, one of the vehicles became stuck. The sun was going down. We pulled him out and the platoon sergeant set some explosives on the dike. The platoon sergeant carried most of the plastic explosives.

When I look back on those days I smile. The Platoon sergeant was one of the older men in our Company. He was 25. By the time we were through getting over the obstacles, it was almost midnight. We called into our captain to tell him the situation. The Captain told our Lieutenant that we would still meet with our rendezvous The Vietnamese would blink flashlights to guide us to their location.

Later I found out that the captain had contacted the ARVNs and they told him that there were too many Vietcong in the area. They could not wait for us. He told them, over the radio, that we were expecting to have the ARVNs spot us with flashlights. The ARVNS said sorry, too dangerous. The captain did not tell us of this conversation.

I was in the lead track, on the left machine gun. We were moving well in the dark and I was ready behind my gun. It was a hot night and the movement of the vehicle across the paddies cooled my skin. Ahead I saw flashes. I swung my gun toward the front. I thought they were gun flashes. The track commander turned and said that it was the ARVNs. They were using flashlights to let us know where they were. I relaxed.

As soon as we got our five tracks stationed around the perimeter, I grabbed the trip flairs. "Its too dark", said the track commander, "just set out the claymore mines." I did as I was told.

We were in a circle with the ARVNs in the center of the circle. In the very middle was a grass house. This was a favorite night station for us. We had become friends with the family that lived in the house and farmed the land. We were not safe but we had stayed here many times.

What I couldn't believe was the ARVNs. They were very noisy. Too noisy. They seemed very busy. Finally, as I was setting out my bed, I couldn't take it any more. I grabbed a rock and threw it at the noisy gooks. Then I made a very exaggerated shhhhhhhh!! I then laid down in my bed and went to sleep. In the middle of the night the water in the rice paddy moved. Probably from the dike we had destroyed. My feet were under water so I moved my bed to high ground and went back to sleep.

Suddenly I awoke with the world on fire. Everything was exploding around me. I jumped up to see a steady stream of machine gun bullets hitting where my bed had been. I ran. As I did, the bullets followed me. The bullets were inches from my heels. I dove into the back of the track as the door closed. Bullets ran off the back of the vehicle. Something was strange but I didn't have time to think about it. I jumped behind my machine gun and began firing. Every flash of light, every thing that moved in the dark received bullets. Suddenly a flash to me left.

I swung around and fired my last bullet from the 100- round belt. I reached back to grab another box and there was an explosion. My arm was almost torn from my shoulder. I felt like I had put my arm in a fire. A recoilless rifle round had hit the ground in front of the track. The driver began to scream that he was hit. I couldn't help. I had to fire the gun. But if I didn't take his place and get the vehicle moving, we were sitting ducks. The first round had given them the range. I had to pull him out and get the vehicle moving. I jumped up on the track and grabbed my friend under the arms. Bullets were flying around me. I tried to ignore them and think about lifting. The driver looked up at me and his face was red with blood. A small head wound bleeds more then any type of wound. This was a bad head wound. I lifted the 175 pound man as if he weighed 15 pounds. I carried him to the back, where he could be safe and jumped back on the top of the track. I climbed into the drivers seat and started the vehicle moving. The 50 caliber machine gun is next to the driver. The gun was firing.

My right ear screamed from the pain of the concussion from the big gun. I ignored it. I looked up at the track commander and yelled that I needed a helmet. He tried to look at me. The round that had hit the driver had thrown small pieces of metal into his eyes. He couldn't see. He turned around to the back and said he needed the helmet. The driver yelled that he couldn't take it off. If he did, his brains would fall out. He was very badly hit in the head. The track commander reached back, felt the helmet and lifted. He quickly spun around and handed it to me. As I put it on I heard the driver scream.

Warm, thick blood ran down my face from the inside of the helmet. I grabbed the cord on the helmet and plugged it into the vehicle.

I hit the intercom. The track commanders voice came over the head set. "Steve, I can't see. I'm just laying down rounds. how am I doing." I looked at the tracers and yelled into the mouth piece that he was shooting too high. He was shooting over their heads. I was driving and firing an M16 that was kept for the driver. Suddenly another flash. The world seemed to go into slow motion. All the sounds stopped and I could hear the recoiless rifle round coming straight at my head. I grabbed the seat release and the seat and I dropped. The round went over my head and under the barrel of the 50 caliber. It exploded in the rice paddies.

I yelled, "fire the 50 to the left and for Gods sake, stop firing at the moon. Shoot lower." I could see movement and the enemy running. Why weren't we getting help? I switched the intercom to the radio. The radio was full of screaming. We had just gotten a new guy. He had been here two or three days. The track commander had been hit and the driver had been hit. The new guy had climbed into the drivers chair and put on the helmet. He was yelling into the radio that we were all going to die. He said that his track commander had both his legs shot off and someone had to help. He completely tied up the radio. We could not talk to each other nor could we get help.

Another huge explosion to my left. I saw the Platoon Sergeants track explode. They had landed a recoilless rifle round into the track. This was the track that held the explosives.

To my right I saw another track on fire. Men were running around trying to shoot and put out the fire. I looked to my back and saw the left machine gunner shooting into our perimeter. I yelled, "You are going to shoot some of the ARVNS". He yelled back, " They are not ARVNs! They have been shooting at us."

And suddenly, No one was shooting at us. The movement stopped and the sky started to lighten. Also the ARVNs were gone. After so much action, I felt like the world had stopped. A bird started to make noise. And over the hill came the second platoon in their tracks.

I climbed out of the driver's seat to check the driver. He was awake and his brains had not fallen out. He was bloody from his waste up. He smiled but didn't say anything. I went to the radio and turned it on. The kid had finally gotten off the radio. I asked for a medic. The answer was that everybody needed a medic. I looked at my arm and saw I was also bleeding. I noticed that I couldn't hear in my right ear.

The track commander was just starting to see again. His eyes were bright red and, when I looked close, I could see small pieces of metal in his eyes. I told him to leave his eyes alone. I grabbed a canteen of water and poured it into each eye.

One of the soldiers from the second platoon ran up with an enemy machine gun. He handed it to me and I had my picture taken. Later I would be amazed at the picture. I looked very old. I opened the gun and took out five bullets.

They laid the dead bodies on the ground and wrapped them in body bags. They would be picked up by a chopper after the critically wounded were gone. The rest of us could go to the hospital in our leisure.

I looked at the dead before they were taken away. The platoon sergeant had been six feet five inches tall. What was left of him fit into a small body bag. He was very badly burned. His face had a fixed scream. He had shown a picture to me of his little girl the day before. She had just been born and he had never seen her.

The Captain had known the ARVNs were not going to meet with us. He had not told us. He also had been told there were enemy in the area. He had not said a thing. We all knew we were always being monitored by the enemy. He knew this, yet he let out critical information that had cost the lives of my friends.

The next day we were called into a formation. Medals were given for bravery. I received, along with most of my platoon, a purple heart. The Captain received a bronze star. After the formation was over, one of the black soldiers from the second platoon walked over to us. His best friend had been the man that had his legs blown off. They always hung around together when the platoons were together. They had been friends when they had been stationed in Germany. He said that he was very angry at the Captain. He wanted to kill him. He had found an enemy rifle in the bushes after the battle and had not turned it in. He had one problem, no bullets. I reached into my pocket and pulled out the five bullets.

It started with one man and soon the entire company was involved. We would all say there had been a sniper. The captain would be shot with an enemy bullet. We would all cover for each other.

Somehow, they found out about it and the Captain was taken away. For a long time members of our company kept track of the captain. He went to Germany and then back to the United States. He was not from the West. I was relieved and angered at this news. We promised each other that, if any of us ever had a chance, we would kill the captain.

It is a big world and there were only a few of us. A man can go an entire lifetime without seeing someone from his past. I doubt that he was ever killed.

But inside of me is a hatred and an anger. To promise that you will murder another human being if you get a chance, and mean it, changes a person. It means you must carry a strong hatred inside. A hatred that can never be released. It turns from the incompetent captain that wanted one thing, Medals, to other incompetent leaders. To me their stupid decisions are a matter of life and death.

Today, I have forgotten the captains name and his face. I must. Today, I can not carry this hatred any more. It is not healthy.

CHAPTER 8

CHILDREN

At the age of 20, traveling to foreign countries did not appeal to me. My only experience out of the United States was to visit Canada. Traveling where people didn't speak my language did not look fun.

After my first week in Vietnam, I found that I was very interested in these people. I thought they were very attractive, especially the women and children. We are used to such a mixture of people in our country that I was amazed at how similar to each other the people looked. We have people of all sizes and colors. They were very much alike. The adults look like children until they reach an age of about 35. Then they look ancient, like the statues we used to buy in China town.

On one hot sunny day in September, we were staying on the outskirts of town. We had set up our camp around an old house that had been built by the French. It was rice harvest time and the Vietnamese would bring their cut rice and pound the rice out of the straw on the concrete patio of the old stucco house.

As I was watching this process, I heard yelling and crying coming from the other side of the house near the well. We used this well for bathing water. I went to investigate. The yelling was coming from a 9 year old boy. These kids usually hung around our camp during the day. I was amazed because they could all speak Vietnamese, Korean, French and English.

Four soldiers had taken the hand sickle from the boy and hung it in a tree. They would not let him get near the tree. They were laughing. The boy was yelling, "Prease GI, prease give me back my knife, Papason whip me if I loose it. Prease give. The four soldiers had been in Vietnam for many months. They looked like moving scarecrows. They were tan and skinny. Their cloths didn't fit and they had a strange look in their eyes. Many months later, when I weighed 155 pounds and had been in combat, I would recognize the scarecrow in myself. But at that time, I had just completed 3 weeks in paratrooper school. I weighed 185 and was in the best shape of my life.

I walked into the group and pulled the sickle from the tree. I handed it to the boy. Then I turned to the four GI's. One said to me, "What do you think you are doing?" I answered that I didn't think they had a right to treat this boy that way. They looked at me like I was stupid. Another answered, "A boy! He's a gook!" and I answered, "He is a Human Being" and walked away with the boy following me.

I apologized for the way they had treated him. After that he followed me around every day. I would give him money and he would get pineapples or coconuts or any thing else I wanted. I have always enjoyed being around kids and he was a good kid.

One day, a few weeks later, he did something that haunted me for many, many years. We were getting ready to move our camp. He took me away from the other people and he drew something on the ground. It was a long tube with something attached to one end. He said very quietly, "B-40, I show you many are hidden. I had been in country such a short time that I didn't have a clue what he was showing me. He said, "Because we friends, no want you hurt. I show you where B-40 hidden. It bery dangerous. So we be careful. I still didn't get it.

Suddenly another soldier ran up to me and yelled that we were moving out. Before I could show him the drawing on the ground, the boy had scuffed it out with his foot. I ran to my Armored Personal Carrier and we were gone.

We didn't return to that area for many months. When we did the boy was gone and I didn't see him again.

About a month later we were in combat. I had a chance to see a B-40 rocket launcher. A B-40 rocket launcher is the Chinese equivalent to our bazooka. It is very powerful and one of the favorite weapons of the North Vietnamese army. My best friend died from one and I still carry a piece of one in my arm.

The boy risked his life to tell me where they kept the B-40's hidden. If I would have known what he was drawing, we could have captured the weapons. In my dream I go with the boy and recover the B-40s. Because I find these, Joe isn't killed. Then at the last minute, I am told that it is only a dream and Joe is dead.

As long as I can remember, I've enjoyed kids. I think it is because my mother and father respected children and enjoyed us when I was growing up. Being the oldest of four children, I was also used to being in charge of the younger kids. Except when I was at the farm. Then I was second in command.

Every year my brother and I got to stay a few weeks at my Grandparents farm. This was one of the greatest times of my life. The dairy farm was always fun. My four cousins also stayed on the farm. Dave and I were jealous because they got to stay all summer.

A typical day started out with a breakfast of oatmeal. Grandpa would make a huge pot of oats and what we didn't eat was given to his dogs. The oats were covered with fresh whole milk that was very good.

When we were finished, the six of us would walk out into the sunshine and Sharon would always ask, "What do you want to do.? Sharon was 10 and the oldest. She always asked me because we were "guests" and they "lived" on the farm. Only I was allowed to answer. The rest of the group followed us. Pam second oldest but she was not the oldest boy! Pam was 9, I was 8, Linda was 7, Dave, my brother was 5 and Jimmy was 4. Sharon and I were in charge. Sharon was more in charge. I would answer with some great idea that I thought would be fun. Sharon's answer was, nah, That isn't much fun, let's go to the barn and play on the swing. I don't remember Sharon ever taking my suggestion.

So off we went to the barn. The barn was half full of hay. There was a long rope that hung from the rafters and we would swing and climb in the hay. That was lots of fun but after awhile we decided it was too hot and the hay was starting to itch. We walked out of the barn and over to the sprinklers. None of us wore shoes or any cloths except bathing suits. The only time we would put on shoes was Sunday when Grandma would take us to town to go to church. Then I had to put on jeans and tennis shoes. The girls wore dresses and they even combed their hair, sometimes.

Sharon was the most beautiful girl I have ever seen. My first memory of her was when I was 5. I was in love with her, as much as a boy can be, from the start. She always treated me special, even when we were teenagers. At night, under the stars on the farm, she would sing to me. I can almost hear her voice. But, above all, she was always in charge.

After we ran through the sprinklers, Sharon decided we should go to the golden apple tree. This was a special tree. It was not located with the other apple trees and it had huge green apples. Sharon said that the apples didn't turn red when they were ripe but yellow. I wouldn't know because we ate them when they were green.

My job as the oldest boy was to climb the tree and knock down the apples. I didn't mind this job except all the best apples were eaten by the time I got down from the tree. Sharon saved the biggest and the best for me. At that age, I didn't feel adult love but something very different. I hated most girls, but Sharon was very special.

We were sitting under a tree near the golden apple tree enjoying the shade. On the hillside was a prairie dog town. Each time one of the rodents popped up his head, Dave or I would throw a rock. This made them let out a loud squeeek and down they would go.

Then Dave picked up a limb from the tree we were sitting under. The center of the limb was a spongy material. He took a piece of it and made a pipe. He stuck this in his teeth and said he was Popeye. We all had to make pipes then, even Jimmy. Then I made a doll. I used twigs to make the arms and legs and the spongy stuff for the body, the head and the feet. We all made dolls. Soon we were busy playing house with the dolls. Of course Sharon had the mother doll and I had the father doll. It didn't take long for this to get very boring for me. The girls were having lots of fun.

I walked over to the creek. It was very hot and the water was moving very slowly. The damp mud was covered with mint bushes. It smelled wonderful. I was walking through it when suddenly the mud was almost up to my waste. I yelled and tried to pull myself out. Soon I was covered with the cool mud. Dave ran over and jumped into the mud hole. I kicked mud on him and we started to throw mud at each other. Sharon saw that we were having fun so she joined us. We found that if we worked our feet up and down we could make the mud deeper. All the other kids joined us except Jimmy. We were having a great time.

Suddenly Jimmy ran as fast as his four year old legs could carry him. He ran straight at Sharon. He stepped into the mud near us and he disappeared into a mud hole.

Sharon and I both grabbed Him by the hair under the mud and pulled. He was stuck. We pulled harder and he came out screaming. He was really mad that we had pulled his hair. He was going to tell grandma and we were going to get a licking. He was covered with mud and we started laughing and calling him a cry baby. Teasing Jimmy was always fun. He was such a brat most of the time.

Sharon said we should go to the house. I didn't want to go because I was sure Grandma would be mad. We were covered with mud that was rapidly drying. I wanted to go back into the sprinklers but Sharon lead us to the house. She was always in charge!

When we got to the house, Grandma came running out. She wore a printed dress and an apron. Her hair was gray and silver and she kept it in a bun. I was taller then she was. She laughed and laughed and kept saying, "Just look at you! O.K. get in the shower. Girls first".

The girls always went first. I didn't mind except the mud had dried and was starting to feel uncomfortable. The three girls went into the shower and we had to wait. They took a long time. Finally Jimmy couldn't wait any longer. He just ran behind the curtain. The girls started screaming for Grandma. The shower went off and one by one the girls came out. First Linda. She had a towel around herself but still looked like a boy. Linda always had her hair short and was the tom boy of the bunch. Then Pam.

Pam had lighter skin and didn't get a good tan. She accidently dropped her towel. She was always doing that stuff. Both Dave and I looked away. Then came Sharon. She had a towel around herself and a towel around her hair. She looked like a young Cleopatra.

The shower started and Jimmy yelped because it was too hot. Dave and I charged into the shower. Then Jimmy wouldn't wash his hair nor take off his swimming trunks. By the time I was through with him, the water was getting cold.

When we got to the Kitchen, Grandma was taking out the bread she had baked. Nothing in the world smelled as good as Grandmas bread. The crust was hard and the center was soft. The smell was wonderful. As she cut, steam came out of the bread. Grandma cut slices almost an Inch thick. We covered the slices with butter and got a glass of cold whole milk. I was in heaven. Sharon said this was real butter.. Not margarine like we got at home. I told her that my mom always bought butter. Not her mom. I guessed that they had margarine because they were poor, but I didn't say so. Sharon and the rest of the kids wouldn't eat margarine because Grandpa said it came from dead Cows that were boiled down. They mixed the grease with yellow food coloring and that was how the made margarine. If Grandpa said it and Sharon believed it, it was true.

In Vietnam we didn't have butter. The milk was called "sterilized milk" and tasted like iodine. I would dream of those wonderful days on the farm and how good that cold milk "right from the cow" tasted.

After I had been in Vietnam about 10 months, my point of view had changed toward the Vietnamese. Maybe too many of my friends had died and maybe because we couldn't tell the friends from the enemy, but I began to hate those people. One of our jobs in the last months was to guard bridges. During the day we would check all the people that crossed the bridge to be sure they had the proper paperwork and to check through their things to see if they were caring weapons. The Vietnamese soldiers were automatically let through if they had a yellow bandanna around their neck. After one battle, four of the enemy soldiers were killed. Each had a small back pack and in each pack was a yellow bandanna.

When it was my turn to check through the belongings of the people and to check paperwork, I told my fellow soldier that we were going to have a bridge closing. For one half hour we weren't going to let any one cross our bridge. This was a heavily used bridge so it didn't make the local people happy. At first the people argued with us but when I aimed my loaded M-16 at them, they went back on the road and sat down. Eventually there were about 50 people waiting. A boy about 12 approached with a bicycle. He had to get somewhere and he wouldn't take no for an answer. He waved his paperwork in my face and talked very loud. I kept yelling, "The bridge is closed". Suddenly I felt a rage growing inside me. I was angry at these people. I grabbed the bike from the boy and raised it over my head. I threw it off the bridge. It fell about 50 feet. Everyone seemed to hold their breath. I yelled that he was lucky because I wanted to shoot him. One of the FNG'S (this was a slang word for the soldiers that had been in country for less then a month. NG stood for New Guy) walked up to me and said things were getting out of hand. He said I shouldn't have thrown the bike off the bridge. I yelled it was my bridge and the kid was a Stupid Gook. He looked at me and said, "No, He is a Human Being".

CHAPTER 9

THE SURVIVALIST

I have been called a survivalist. There are many reasons I feel this way. As a 44 year old male that has lived in a war, I have learned that we are not as safe as we believe.

I was 21 in Vietnam. I remember writing to my mother and trying to describe how it felt in a war. The strongest feeling was the loss of security. I keep my rifle close at all times and I had to rely on my fellow soldiers to stay alive.

At home, in the United States, we have a circle around us of protectors. The circle is very large and we do not have to protect ourselves. In the war, this circle of security is very small. The only real survival is you and the guard on duty.

I have often felt that this security could be broken. We, as a country have lived so long feeling safe, we can't imagine any other way.

One other lesson the army taught was that the government is not as wonderful as we think. Our government could easily be controlled by another government.

So, I am called a survivalist. I believe it is possible for us to loose our freedom. In the extremes of my imagination, I have built a fantasy.

In this fantasy, our government is overrun. We have one day to prepare. Also, we are taken over, not destroyed. What would you do if you had one day to escape? One day to set up a resistance force? Perhaps this enemy is the South? What would we call the resistance force? I thought Viet Cong was a good name. It worked for another country.

I would see the last day as a day of looting and chaos. A day when everyone would run to the grocery stores and take everything. The most popular item would be a rifle, assuming they are not outlawed.

I would go to the sporting goods store and pick up a backpack and all the freeze dried food I could carry. I would get camping supplies and archery equipment. Why not a rifle? Because it is heavy. Because the bullets would eventually run out. And, I could build new arrows.

The food should last until I become successful at hunting and finding food. I would not expect to be alone. This would be the beginning of the resistance force.

To learn how to be a survivalist, I bought a bow and some arrows. I have studied edible plants and I began camping in the woods. I tried hunting. I also took my bow to a lake and shot fish (carp). Eventually I even shot a grouse and a pheasant with the bow. The pheasant I shot in the air. This is not an efficient way to hunt. I lost too many arrows.

Hunting is not as easy as it looks. Especially with a bow. I had seen all the TV specials where Bambi is walking around taking food from peoples hands. Believe me, it is not that easy. To hunt with a bow takes great skill.

Then came an opportunity of a life time. I was asked if I wanted to hunt with two of my uncles. We hunted elk. My uncles are very experienced hunters.

In my twenties, I had gone on one hunt. This was a hunt for men. I had been in Vietnam less then a month and had not yet been in combat. The army fills the heads of young men with ideas that make them think they will live forever. They teach there is no greater glory then to be in combat. This brain washing is good for the army because the more aggressive the troops, the better chance to win the battle. On the other hand, it kills a lot of people.

We were told that a Viet Cong camp had been located and our Captain asked if there were volunteers that would like to "kill a few Charlies?" I was ready. I was one of the first to raise my hand.

To prepare, I threw two bandoliers of M16 rounds over my shoulders and attached a canteen of water to my belt. I was ready. I was carrying over 400 rounds of ammunition.

We were a group of 20 plus hungry fighters. That is the way we looked standing at the bottom to the hill in the 120 degree heat. We were all sent back to get our helmets. The sergeant said that we should look like soldiers.

We started up the hill. Immediately, I realized this was not going to be fun. The underbrush was very thick and the hill was steep. The only easy way up the hill was along a dry stream bed that was probably a water fall in the rainy season.

The sergeant in charge yelled out, "Where is that new kid. What is his name? Manthey."

I yelled, "over here". I was holding on to a bush and trying not to stick my rifle barrel into the dirt. My helmet kept falling off.

He told me to take the right flank. I looked to the right. It was solid brush. To say it was hot is an understatement. Sweat was always running into my eyes and my shirt was soaked. I spent more time on the ground then walking. My goal was to try to keep up.

It was easy locating the rest of my squad. They were very noisy. With the yelling back and forth and the crashing around in the brush it was louder then a bull dozer cutting a road.

I also could smell the many cigarettes they were smoking. I have always been amazed that when people need oxygen the most, they smoke the most.

I took the bullets out of my rifle. I had too. I spent so much time falling I was sure I would accidentally pull the trigger.

Suddenly someone yelled that there was something ahead. I put a bullet in the chamber.

I think many people were killed in Vietnam because they weren't prepared for the war. The first M16 rifle I saw was the one assigned to me in Vietnam. They gave us a one week survival course when we arrived. This was taught by troops that had been in Vietnam for at least one year. The first thing they said was to forget all the crap we had learned in "the world". In Nam we called the United States "The World" I soon realize that they were right.

So now I am called a survivalist. The hunt was very different with Uncle Jim and Uncle Howard. They knew the country and the elk. They understood what was needed and were prepared. We spotted and hunted elk all day. At one point, Uncle Howard saw some tracks on the side of the road. He said there were elk nearby. I studied the tracks carefully.

As we moved into the woods, he told us to move carefully and be very quiet. He tracked the elk. I was amazed. I looked at the ground and saw the forest floor. He saw elk tracks. Suddenly he stopped and started to carefully look to our left. I looked too.

He whispered, "elk and nodded his head at the trees and brush ahead. I saw trees and brush.

One moved! Then another and suddenly the beautiful animals were all lined up about 150 yards ahead. Almost as soon as they appeared they were gone. We followed them for awhile until Uncle Howard said they had crossed the river and he knew exactly where they went. He said they would be waiting for us. We went back to the truck.

After driving around for awhile we crossed the river and drove down an old dirt road. Trees leaned over the road and the truck pushed the limbs out of the way with the windshield. When we got to the river, it was determined that Uncle Howard and I would follow the elk and the other members of the group would go with Uncle Jim. We would try to send the elk to Uncle Jim's group.

We followed the river for awhile until Uncle Howard picked up the tracks. He would point the tracks out to me. I tried very hard to be a good student. Uncle Howard moved through the brush like it was easy. The leaves moved and I know he touched the ground, but he was absolutely silent. I felt like my feet were too big. Each time I would drag my feet over a fallen log or break a twig, he would look back at me. He didn't smile. Most of the time I held my breath and walked on my tip toes.

Suddenly, I heard running ahead. I touched Uncle Howards shoulder. He stood up and signaled me to follow. He took off running.

The army does not teach soldiers to be quiet, to move through the brush like a deer. They teach that you have superior fire power.

When in doubt, call in an air strike. The troops on the ground are used to finding the enemy so the choppers and artillery can blow them up. This is a good concept when you look at the big picture. For the soldier on the ground, this is not healthy.

As a young soldier on my first ground mission, I didn't understand this. I was ready for a kill. After climbing that stupid hill for at least an hour, the call from the troops that something is ahead, was music to my ears.

As I started through the jungle I suddenly came to a clearing. It looked like someone had sat in the clearing and when I moved to the right place, I could see far into the rice paddies below. I could see our company of armored personnel carriers and I could even see our soldiers walking around. With little effort I could easily shoot any of them. It occurred to me how silly they looked. The vehicles were in a circle like the covered wagons. In the center of the circle was the command track where the captain stayed. I could even see him talking to one of our lieutenants. It would not take a rocket scientist to determine which was the command track and even who was giving the orders. I wondered how many Viet Cong had sat on this very spot and aimed their rifles at our captain. Then I looked at my track and wondered if they had ever aimed at me.

So far, I had not heard shooting but the yelling and the noise became louder. I finally joined the group.

We had walked into the camp of the enemy. It was in a ravine about 8 feet deep. The ground was steep enough that the Viet Cong had dug sleeping areas in the hillsides. There were about 15 dug out areas. This was smart because they could sleep and not have to worry about artillery rounds. The rounds would explode in the trees overhead. The caves were deep enough so the sleeping men would be safe. In the center of the area was a fire and a pot for tea. The pot was still boiling! This was in the trees.

Everyone was running around looking for souvenirs. Not me, I was afraid that anything worth having would be mined. I didn't have a lot of combat training but I wanted to go home and I didn't want to take stupid chances. The sergeant came over to me and asked if I had seen the enemy. No I answered. I had guarded the flank as he had ordered. He said look. I looked on the ground and it looked like car tire tracks. The Vietnamese made sandals from old car tires. I had seen them many times. The tracks lead toward our right flank. At the edge of the ravine, The tracks were covered with army boot tracks. My boot tracks.

The sergeant looked at me and asked again, "Are you sure you didn't see anything?"

The image that filled my mind was 15 or more Viet Cong hiding in the brush as one clumsy soldier stumbled within feet of there hiding places. They would not have fired if they thought there was a chance of alerting the horde of Americans that had walked into their camp. They would not have seen how many men were coming but I'm sure they thought there were many, We were so noisy. I also imagine there leader deciding they wouldn't shoot because, If they waited, I would probably kill myself falling down anyway.

Very disappointed that I had missed my big chance to kill an enemy, I joined the rest of the soldiers for the trip back down the hill. I was very hot and had long ago drank all my water.

About 10 minutes into the hike, we stopped. I asked why and was told that many of the soldiers had dropped most of their heavy equipment, including ammunition on the trip up and they had to find it. Some was never found.

Wnen Uncle Howard took off running I followed him. Even running, he made very little noise. We ran across a flat area and up a short hill. As we came to the top of the hill I looked to my left and there was an elk. It was looking at me as if wondering what I would do next. I was so excited, I drew my bow and forgot everything I had learned about shooting. The elk was about 25 yards away and was beautiful. The color is amazing because, in the forest, it blends so well with the background that you have the impression there isn't an edge to the silhouette of the elk. I fired. I fired without aiming. The arrow went over the elks head and into the trees. I pulled out another arrow just in time to see another elk, or was it the same elk in a different location? standing 35 yards away. This time I was calm. I anchored the arrow against my cheek and aimed behind the shoulder, where the heart and lungs are located. I released and the arrow went into the side of the elk, clear to the feathers. The elk ran.

Uncle Howard tracked the elk and eventually found it. It was a clean kill. When we got to the elk Uncle Howard leaned back and let out a very loud whoop. I was so excited I also let out a loud war cry. He looked at me and asked what I was doing. I smiled and said the same thing he was doing, expressing my excitement. He said that he was calling to Uncle Jim so he could help us. He said something else but I didn't hear it.

I doubt that all of the troops in Vietnam were as poorly prepared as we were. We were tank troops and not trained ground search and destroy missions. In the year I was in Vietnam, we didn't leave our vehicles again. Probably because we were so busy with our own type of war. I do not know if the infantry were as poorly trained as the armored divisions. If they were, we would have lost thousands of men.

I do believe that the men in World War One and Two were better troops then the men in Vietnam. Why? Because they grew up hunting and living on farms. I was amazed in basic training because many soldiers had never fired a rifle. Remember the old movie about Sergeant York? When he made the turkey call so the German soldiers would look up? He didn't learn to hunt turkeys living in a city.

I'm not really a survivalist. Friends have called me that but I think a survivalist would live in the woods. I'm just a person that works and lives in a city. If my fantasy came true, I don't really know if I would survive. I do know that I would survive better then many of my friends. How? I would go find my Uncle Jim and my Uncle Howard.

CHAPTER 10

CRAZY PEOPLE

Within the realm of Normalcy, there are individuals that have a different or unique reality. They see what we see but they respond differently. These differences are readily noticeable to the normal population. The different people get many different labels. To me, the best label is crazy.

It was the summer before my third grade and a very different time in my life. A major change had occurred. We had moved to our new house. It was BRAND NEW, even the streets were not built. My brother and I had met all the neighbor kids and were anxiously awaiting the new owners of the houses that had not yet been completed. To our excitement, a moving van was in the driveway to the house next door. Dave, three years younger then me, and I went to greet our new neighbors.

The first person we met was Margaret. I immediately hated her. She was three years older and a snob. As we tried to communicate with this snotty kid, her sister came around the corner. Her name was Linda. She was one year older then me. She was carrying a book and was overweight. She was nicer, but we were disappointed. No boys.

Dave asked, "Are there any other kids in your family?" Margaret answered that she had another sister and a brother. She told us that her youngest sister had been locked in a closet and that she was "different". Dave looked at me and said one word, "Crazy". We both knew what he meant.

Barbara joined us. She was one year younger then me and she was crying and screaming about wanting something in her room. Margaret said, "I told you she was nuts". Margaret told her to leave but she stamped her feet and started yelling for her mom. I picked up a rock and casually threw it into a mud puddle next to their house. I did it hard enough to splash mud on Barbara. She let out a scream for her mom and ran away. Dave and I laughed. Unfortunately I also splashed on Margaret. Margaret said that we splashed mud on their new house and she was going to tell her Dad. I was in big trouble. She didn't tell. She held it over my head for almost five years. Every time she wanted something, she would threaten to tell her Dad about the mud on the side of the house. I hated Margaret.

Because Barbara was crazy, Dave could do anything to her he wanted. She would walk out of her house and he would stop whatever he was doing to get on his bike and chase her. Barbara would scream and run home. Then she would stand on the porch and yell that she was going to tell our mother. We would laugh and go back to whatever we were doing. Dave would set up and chase her blocks on his bike. As he would be right behind her, he would be making car engine sounds. We used to believe that Barbara had problems but our job was to drive her crazy. When Barbara was about 13 she came to our house and announced to my mother that she had gotten her first bra. After my mother left, Dave and I said that she was a liar. She said, "Oh yeh", and lifted her blouse to show us. Dave said that she was crazy and left. I said, "big deal".

Crazy People   67

It was just a fancy T shirt and she didn't need a bra. With a strange look in her eye she looked at me and lifted her bra. I guess my eyes got real big because she laughed and turned her back. "Wow", I remember thinking, "Too bad all girls aren't crazy, and too bad Barbara was so fat!"

One day Dave and I were shooting our BB guns out our bedroom window. We were trying to shoot birds off the telephone wires. Barbara walked out of her house and up the sidewalk. I said to Dave, "watch this". When she was next to a telephone pole I shot my BB gun. It hit the pole and Barbara jumped. Causally she started up the sidewalk and when she was buy the pole, I fired again. This time she heard Dave and I laughing and she yelled as loud as she could that she knew we were shooting at her and she was going to call the police. At that, Dave and I both fired near her and I yelled that she had better get in her house or we would shoot her. She screamed and ran home.

About ten minutes later a police car arrived and Barbara met them in her driveway. Dave and I were looking out the window when Barbara pointed at us and the policeman said very loudly, "O.K. You two come down her right now". Dave was half way under the bed by the time the policeman stopped talking.

I grabbed him and we went downstairs. The policeman asked if we were firing loaded rifles at this girl and if we had threatened her. I answered that we were shooting our BB guns at targets in our back yard and that Barbara may have heard the shooting and thought we were shooting at her. Tears filled her eyes. Dave looked at the policeman and said, "She is Crazy". Using my most grown up voice I explained that everybody in the neighborhood knew that Barbara was "a little slow". That we didn't do anything. I turned to Barbara and said, as if talking to a two year old child, "Do you understand how much trouble you can get us into by lying. Barbara you better tell the policeman you made a mistake." By know the tears were streaming down her face. The policeman turned to Barbara and she started to shake. He asked her if she wanted to press charges. She was crying so hard now that we could hardly understand her. She answered, "No". The policeman said that she could get in a lot of trouble by taking up his time making false charges. He was very stern with her. She turned and ran into the house crying for her mother.

Dave and I walked into our house and, as soon as the door was closed, burst out laughing. Finally, Dave said that it was a good thing she was crazy. That night at the dinner table my younger sister, Debby announced that Barbara had called the cops because Dave and I were shooting our BB guns at her. Dave and I told the same story that we had told the police and ended it with "besides, she is crazy". Debby loudly exclaimed that she was crazy because Dave always chased her with his bike. Mom was angry and asked if we were trying to drive that poor girl crazy. "No mom", we answered. "We can't help it if she is crazy." Both Dave and I knew that we were driving her crazy. She started out slow because she had been locked in a closet but we were driving her crazy. We never doubted that we drove her crazy.

My first lesson: A traumatic experience weakens a person. This lets the door open so that one or more people can drive a person crazy.

Up the street and around the corner lived clint and Bobby. We didn't know how old they were. Mom said they were retarded. We just labeled them Crazy. They were always dirty and were covered with sores, Bobby was the youngest. Every once in awhile they would roam the neighborhood. I didn't like to be around them but Dave loved to watch them. What he liked the best was that they were always hitting each other. Not just hitting like kids do. They would take two by fours and hit each other in the head. Dave said they were like the three stooges and he thought they were real funny. They liked to take off their pants and chase my sisters into the house. Dave especially liked to see them chase Barbara.

The neighborhood kids said that they were crazy because their mother and father were brother and sister. Dave and I didn't believe this because we thought the grossest thing in the world would be to marry your sister. When we heard this we looked at our sisters and said , "YUK!"

So, that evening, we went to Clint and Bobby's house to see the parents. We wanted to see if they were really related. We road our bikes around the street waiting to see the parents arrive. When they did arrive we were greeted by an overweight couple that looked very tired. They drove an older car and, to our surprise, were carrying a small boy and a baby girl. We weren't real sure but we determined that both younger children were also nuts. I did not think the couple looked like brother and sister but Dave was sure of it. Right there and then we vowed never to marry our sisters!

My mother was president of the PTA. This was a very big job and we were proud of her. When they announced that a PTA meeting was being held, I would announce that my mother would be there because she was the president. One evening my mother was very upset, Clint and Bobby's mother had applied to have Clint put into the first grade. She said he needed an education and that he was capable of being in the public school. My mother was on the phone all evening talking to the principal, teachers and other mothers. All agreed on two things: Clint should not be allowed to go to the public school and my mother should tell the poor woman.

My mother felt very sorry for Clints mother and I remember how sad she was after she told her that Clint could not be accepted in school. Everyone patted my mother on the back for doing the right thing but my mother was very upset and was not entirely sure she had done the right thing. My mother had a lot of compassion for other people.

Some time after that we heard that Clint and Bobby had found a gun and had killed each other. My brother was very sure that the parents had shot the kids to get away from them. Then less then a year later there was a fire in the house and the little boy died. Now Dave was convinced there was foul play. After that, the family moved out of our neighborhood.

My second lesson: Being crazy is very dangerous. When I was in Vietnam I had many traumatic experiences. The Army was always manipulating me and I was convinced that they were trying to drive me crazy. Everything seemed so unreal. All the morals and important rules I had grown up with did not apply. Killing was good. Hurting people was good. Paying for sex was OK. The most important issue was that your friends were always right. No matter what they did, it was acceptable. You covered there backs and they covered yours. Survival at any cost. That was the only law.

I was the driver of an armored personnel carrier. We drove those vehicles very hard. They had a V-6 Jimmy diesel engine and we beat them. I have a lot of respect for that engine. Of course, eventually they broke. When this happened the limping vehicle was dangerous. I started noticing my vehicle was loosing power. It also began to use a lot of oil. I requested that it be taken to the rear and an new engine installed. My request was granted.

This meant that, for one night, we got to stay in a rear area. The four of us parked the track and ran to the NCO club. We were getting pretty drunk when the floor show began. It was a Korean band that could only play one good song: "we got to get out of this place". By ten that night we were real real drunk. My Polish friend asked if I wanted to go to LaBa. I didn't know what LaBa was. He took me to our vehicle and we pulled out two cases of Crations. We threw the heavy boxes on our shoulders and started staggering down the dirt road. We turned left past a sign that said "off limits". He said not to worry.

We came to a guard tower and were greeted with a "Halt! Who goes there". My friend laughed and laughed. He said we are American Soldiers and nobody in Vietnam said " Halt! Who goes there" Then the man in the tower yelled," Stop or I'll shoot This made both of us laugh and laugh. We picked up our boxes and walked past the guard tower. We soon came to a second tower. My friend was starting to sober up a little. When the guard yelled Halt, Who goes there, Ski was not in a laughing mood any more. He told the guard where he could put it. The guard yelled that he was going to contact his superiors, that he had orders to shoot us if we didn't obey him and that if we went any further we would be court marshaled. I told him to go to hell and we walked into the town of Laba.

LaBa was the most wonderful place on earth. It was a small town on the edge of the South China Sea. The streets were sand and cool wind blew through the village all the time. There were no men in the village. The town were only inhabited by women, Young, Beautiful women. I felt like I was in the movie, South Pacific.

I asked my friend why this place was off limits. I suspected that the officers kept this place to themselves. I was wrong. Ski answered that the town was known to be used by the Vietcong as an R and R Center. I was suddenly glad that I had brought my rifle. Of course I didn't go anywhere in Vietnam without it. Ski had only brought a pocket full of grenades. I wished I had more then three clips.

I forgot all about the Vietcong when the young women saw us. They came running out of the houses and gathered around us. There were about 15 very attractive women. Ski turned to one and said that he would give them a case of Crations if he could stay the entire night with one of the women. They agreed. I picked out a very, very attractive woman. She led me to her house. I could look out the window and see the waves hitting the beach. The house had sand floors and cardboard walls and corrugated tin roof. What attracted me the most was the bed. It was a real bed with pillows. I had not seen a pillow in Vietnam.

We walked out of the village the next morning without any incidents. I had expected trouble with the guards but we were in luck. I got back to my repaired vehicle and drove back to combat.

A few months later I had to see a dentist. I needed a tooth pulled. I grabbed my rifle and hitch hiked to the rear area. When I checked in at our base, I was told to go immediately to the dentist. I arrived and was put into a chair. The dentist gave me a shot to numb the tooth and stepped out. When he came back, he asked me to step down from the chair and sit in the waiting area. I was freezing. The dentist had air conditioning and I was used to 100 degree weather. At night it would sometimes get as low as 85 degrees. I asked what was the problem. The dentist told me that an officer had arrived and that he needed his teeth cleaned. I was angry and told the dentist that some rear person that never saw combat was pushing me aside because he was an officer. He told me to sit down and shut up.

I sat long enough to have the numbness wear off. When I was back in the chair I informed the dentist that I was no longer numb. He said that I had better be tough. If another officer came in, I would have to wait. It is better to get it over. I was very angry. The dentist climbed up the chair and put both knees on my chest, inserted a tool that looked like pliers and pulled. He was pulling out a wisdom tooth. His pliers slipped and he almost fell off the chair. In the process he cut my lip and it started to bleed. He climbed back up and pulled. It hurt-bad!

With a mouth full of gauze I walked to the nco club. I needed. something to ease the pain. The war could wait for another day for me to come back. Besides, if the army was in such a hurry to get me back in the field, The officer would have waited his turn.

IN the NCO club was a group of 18 year old kids. They were very, very drunk. One was crying. Actually he was wailing. He had grown up with his best friend. They had joined the army together. They had come to Vietnam together.

A week ago this soldier had been wounded. Not a bad wound but bad enough to keep him out of the next mission. On that mission his entire squadron had been over run. There had been some hand to hand combat. The squad had walked into an ambush. The kids best friend had been shot in the top of the head. He had nowhere to hide and was laying as flat as he could. The young soldier was very upset that he hadn't been with his best friend. He kept saying that he wanted to die. That he was supposed to die. He should have been there. I will always wonder if he made it home. Keeping alive was hard work and it took a lot of luck. If you wanted to die, Charlie was very happy to grant your wish.

I was feeling less pain but not enjoying the sadness of these people. A fellow paratrooper came into the bar and I introduced myself. As with all soldiers, I asked where he was from. He answered, "Vietnam". I asked what he meant. He said that he had a wife in Vietnam and that he lived here. He enjoyed the war and did not ever want to go back to the United States. He had been in Vietnam so long that they tried to force him to go back to the United states every 6 months. He would fly to Camron Bay and catch a chopper back here. Then he would stay with his wife for the month he was supposed to be in America. I asked what his job was. He was now a cook. Before he had been a grunt in the field but had been wounded so many times they wouldn't let him back into the war. This made him real angry. In some famous battle he had saved a bunch of GI's and had gotten a silver cross. He was easier to listen to then the crying soldier but I knew right away he was Crazy! He also liked to talk and he kept buying me drinks. We stayed for the floor show. This was a band from the Philippines. They had one favorite song that they played many times, "We got to get out of this place". They had girls in their band so we stayed.

About midnight he asked if I wanted to go with him to see his wife. I asked where she lived and he said, "LaBa". I told him I had been there before but going in had not been very much fun. He said he had a way in that wasn't guarded. Only he and the Vietcong knew the way. I said, "sure, I'd like to go".

We walked into the village and he was greeted by the women. He could speak their language and he wanted to buy some dope. I wasn't interested (Mom, this sentence is in here for you).

I looked up the street and saw the woman I had been with before. She looked at me and smiled. Then she looked at him and her face turned white. Suddenly there was much talking with the women. The soldier turned to me and said that he would like me to meet his wife. Then he said that he understood that I had met her before. I gripped my rifle. It had 18 rounds in it. He did not have a rifle. I waited to see what he would do. I wanted to cock the weapon but I was afraid this might cause him to react. He was crazy. He asked me if I had slept with his wife. He said this very slowly and with a lot of control in his voice. I adjusted my feet, positioned my rifle so I could cock it and fire rapidly and then i answered, "yes, but I had no idea she was married."

He turned away from me and the girl. He walked over to one of the houses. He brought his fist down on the metal roof and the structure started to collapse. He stepped forward and started beating the house. It fell. Then he walked across it and knocked down another. His wifestood by him and just watched. One of the women grabbed me and said let us go before he starts doing that to me.

She was tall and more chesty then most Vietnamese women. I guessed I would stay with her. She smiled and said that she would like that. Because of the bad things that had happened, she would not charge me but I had to wait until she was done with her card game. The Vietnamese love to gamble.

I sat in a room alone. I was enjoying just sitting on the bed. It somehow reminded me of home. I heard another soldier on the other side of the wall. I said "hi" and that I was waiting for my date until the card game was over. He said that his date was also in the same game.

I asked him the same questions all soldiers ask. What unit are you in? He told me. I didn't really care. I told him I was with the 173rd. Our nick name was The Herd. He answered that we were all nuts. Our reputation was of a tough group of fighters that got in more combat then the other units. Then I asked the second question," Where are you from?" He said he was born and raised in California I answered that I was born in California but raised in Seattle. If you said Washington everyone thought D.C. He asked what part of California. I answered San Luis Obispo." Oh yeh?" He answered, "I was also born in San Luis Obispo". I asked what day? He answered February 21st. I said that was my birthday! I said I was 21 and he said he was also 21. We laughed about being in beds next to each other again, after all these years. Then his date returned from her game. She led him to another room.

I looked up and there was a girl watching me. When I noticed her, she jumped back. Eventually she came back. I asked her to join me sitting on the bed. She sat down and motioned that she could not speak nor hear. She was quite a flirt so I was enjoying teasing her.

I got up and walked over to the card game. "Are you going to be all night?" I asked. She answered that she was winning and that I should either join in or go look at the water. I asked her who the deaf girl was. She said that the girl was "dinky dow". This is the way they say crazy. They had taken her in because she would have starved otherwise. "She cleans up around here." I tried to understand the card game but it was too complicated. I walked to the back of the house. The back had a roof over the top but was open. It was kind of a porch with a sandy floor. There was a full moon. In Vietnam the moon rises over the water. I sat on a couch and watched the waves and the moon. Off in the distance there was a fire fight going on and I could see the two sets of machine gun tracers shooting at each other. I felt someone touch my back. It felt very gentle and very timid. I turned and looked into the face of the deaf girl. The moon lit up her face and she seemed such a quiet, fragile person trying to survive in this terrible place. By now I could hear the artillery rounds landing at the fire fight. Next the choppers would be involved. The noise seemed to get farther away when I looked at the girl. She put her arms around me.

The next morning I woke up sleeping on the couch. No one was around. I looked for the deaf girl to at least say good by. I wanted to say something. I had been so amazed when she had turned out to be a virgin!

My mouth tasted very bad and I had a great start on a hang over. It was getting hot as I walked down the road back to base. The events of the night had been so bazaar. The crazy person and his wife. Of all those women, what a strange thing that I had been with her. And the soldier that was born on my birthday. I didn't even get to see what he looked like. And the deaf mute virgin. It all began to seem so unreal.

I got back to my base, ate breakfast and caught a chopper back to my buddies. As the chopper took off I was still thinking about the events of that night. I realized that I couldn't tell anyone. Who would believe me? They would think I was crazy! I questioned if those things had really happened. What if I only thought they happened. No they happened! Unless I was going crazy. Then anything could happen, and I could not be sure if it was true. No it happened. Unless, Unless.

Many things happened in Vietnam. Many strange and frightening things. The most frightening thing was the day I questioned my sanity.

CHAPTER 11

CHRISTMAS

Everyone has enjoyed watching Peter Pan. Remember the part where the children were trying to fly? Peter said, "Think of something happy and you will fly and the little boy yelled, "Christmas!!!" and flew away.

That is how I feel about Christmas. Christmas was a time of my father. My mother was wonderful and surrounded us with a great feeling of family and fun and I don't want to diminish her part but, when I think of Christmas, I think of my father. I guess it was because my mother always gave us a good feeling about home and Christmas transformed my father. All year my father squeezed his pennies. He was very concerned about the high cost of living. We heard lectures about how expensive it was to raise 4 children "in this day and age". On Christmas, Dad relaxed and enjoyed giving. Christmas eve 1953.

My brother and I were very excited. We were sure we were getting scooters. We had picked them out of the sears catalog and had asked Santa Clause. We were completely convinced we would get scooters. These weren't just scooters. They looked like rockets and you could pedale them like a bike. Dave was three and I was six.

"OK kids, I think it is time for Santa to come, lets go into the bedroom and read the Christmas story", said my father. This was our tradition. One of my parents would stay out and "help" Santa while the other (usually my father) would read the Christmas story. This could not occur until after dinner and after the dishes were cleared. We heard this story every year and I don't think I ever listened to one word.

Suddenly we heard bells ringing. These were Santa's sleigh bells and signaled Santa leaving. These bells sounded a lot like the bells my mother had gotten from my grandparents farm. Dave jumped off the bed and ran for the door. "Dave, stop right there", said my father. Dave gulped. He wanted out that door but he wasn't going to give my Dad "back talk". Back talk was the worst thing you could give my Dad. "Get back up here and let me finish the story. Dave and I sat there staring at the door. My father slowly finished the story. "OK, lets see what Santa brought". This time we followed Dad out the door. There, in front of the tree was one scooter, just like in the sears catalog. Wait, there is only one!

And then my father went into the kitchen and brought out a new bicycle. "This is for you, Steve". It was a black German made bicycle. It was not the standard bike of the day with balloon tires. It was what we used to call "an English racer".

I don't remember the rest of Christmas. The next day we got up early and went to church. My father commented that it was so much better to open presents on Christmas eve because the children had already opened there presents and weren't so misbehaved in church. Embarrassing your parents by being misbehaved was the worst thing a kid could do in public.

When we got home, Dave and I gave our new toys a walk around the block. The streets were wet. This is typical in Seattle on Christmas day. I did not know how to ride a bike so I pushed it around the block. Each time a car would pass us I would wave. I was showing off my bike. It was a bike that would become part of my identity. It was the first thing I owned that was not a toy or clothes.

Christmas Eve 1960. Now there were four kids and I was 12. This was the year that Dad decided it was stupid to spend "Good Money" on something that would be in our home for only a few weeks and then we would take it down. This was the year that we went to the sears parking lot to get a $1.00 tree. "Dad, these are the ugliest trees I have ever seen. Can't we get a good tree?" I asked. That was when I got the lecture about "spending good money". "Arnold, I'm afraid I think Steve is right", said my mother, "Most of these trees only have three or four limbs. I was hoping for something a little bit fuller." Dad grabbed two trees and pushed them together. "I'll just wire them together and we will have a full tree." Dave whispered to me "I sure hope my friends don't see this." "Me too", I whispered.

As Dad tied the two sad looking trees on the roof of our station wagon he looked very happy. Imagine, He said to all of us, A Christmas tree for only two dollars.

When we got home, Dad had two problems that he had to solve. First, the tree stand could only hold one tree. So, he cut off one of the stems and wired the two trees together with green wire.

The second problem was the star at the top of the tree. It was designed to go on one tree top, not two. So he cut off the top of the other tree.

What do you think", He said as he kept turning the tree, "which is the best side". Mom was trying to make the best of it. If we use lots of tinsel, I think that is the best side. And I think we can cover up the green wire with bulbs I think it will be great. Dave looked at her like she was crazy.

Now it was Christmas eve. Of course, the tree that had its stem cut off had not gotten any water. The needles were turning brown and most of them had fallen off the tree. I do not remember what I got from Santa that year. I know we went upstairs and read the Christmas story with my Dad and I know we had fun opening presents. The one present I remember was from my Grandma.

My mother grew up with a large family. She had eight brothers and one sister. Most of the ten children have at least four children. To me, my grandmother had so many relatives, It was amazing that she even sent me a present. Usually, she would send small gifts and my brothers and sisters would hardly notice what she sent. My fathers parents overwhelmed us with many wonderful gifts, as did my parents.

Grandmas gift to me that year was a book. I opened the present and discarded it immediately. Later I put it above my dresser, and it sat there for years, unopened.

One rainy day, after my grandmother had died, I read the book. As I read it I wondered why she had sent this book to me. Did this book have a secret message for me. Grandma had always seemed very wise. I had spent many hours working on the farm with her during the summers and I enjoyed being with her. She had a very high voice and long gray hair. I think she was under five feet tall.

The book was about Dutch children that had played together. They especially loved to play in a church tower. The dike had broken and they had climbed into the tower and they had many adventures before they were saved. I think my grandmother gave me the book because she wanted a book that would be challenging for me to read, that was about kids my age and it is about other cultures.

I received many Christmas presents as I grew up but this is the only one I still have.

My favorite part of the book is what is written inside the cover. It says, "To Steve, Merry Christmas, Love Grandma and Grandpa".

Christmas Eve 1968. Vietnam. My first Christmas away from home. I have been in Vietnam since August and have not yet seen any real combat. It was not until there was a bombing halt on the Hoe Chi Min trail that I got into combat.

Months of riding around Vietnam on an armored personal carrier. Months of heat and boredom.

"Mail Call!" Yelled the sergeant from the back of the truck that had come out to give us supplies. He started yelling out names. "Manthey", He finally yelled. I jumped up and he handed me a box about three feet long.

The strongest memory I have of Vietnam is the smell. We are used to fresh air and the subtle smell of flowers and trees. In the dry season there is the smell of dust and heat. In the wet season is the smell of mud. The blue mud of rice paddies. There isn't any way to describe the smell. It is something every Vietnam Veteran remembers. It is a foreign awful smell. Christmas is in the wet season.

I opened the box and, inside was a Christmas tree. As soon as I opened it, the smell of home came pouring out. It was a tree from Uncle Howard and his family. In the heat and the long trip from home, all the needles had fallen off the tree. As the wonderful smell poured out of the box, I saw heads turn. The other soldiers started to walk toward me. "Hay, what is that smell", said one soldier. Look, Manthey has a Christmas tree. Who would send a tree. That smell is wonderful." Suddenly I was surrounded by 15 soldiers. All were smiling and smelling the tree from home. I pulled the tree out of the box and we all started laughing. It was about three feet tall and didn't have one needle. Also, inside the box was a box of tinsel and a box of ornaments. I set the tree aside and went into the track. I pulled out an empty amo box and carefully put the needles into it. The amo boxes seal themselves when they are closed.

For over two months, I could open that box and smell home. I went back to the tree and decorated it with the stencil and the ornaments. Then I mounted it on the back of my track. I just got it strapped down when we got orders to leave.

Our orders were to go into the chopper camp and stand down. This was on the edge of town and one of the more primitive camps in Vietnam. We were told that we could not go to town but we didn't have to worry. Charlie had declared a temporary peace during Christmas. We weren't even going to stand guard!

What a gift. To once again feel safe. What does a twenty year old male do with a bunch of other similar age men when they want to celebrate? We went to the club and drank. This was not like the nice clubs in the rear areas. This was a bar with a few bottles of whiskey and gin and a bar tender. Drinks cost one dollar.

We stayed up most of the night and talked about home and Christmas and women and home. At about three o'clock I walked back to my track. Someone was sleeping in my hammock! I looked around and started to get angry. Where was I going to sleep? Two of my fellow soldiers were still awake and I joined them. I could hardly talk. Suddenly one of them turned and threw up.

The smell was all I needed and I also threw up. When we were done the third soldier said merry Christmas, threw me a blanket and went to bed. I got up wind from the place where we had been and laid down on the ground and went to sleep.

Get up, Charlie has broken the truce and we have to be ready. My head hurt. I was not in the mood to play soldier. What happened, I asked. The Captain just came by and said that about a mile away, three American soldiers had been sleeping on guard and had their throats cut. I tried to smile and said that with the way my head felt, that might not be so bad.

My best friend, Joe walked up to me and asked where I had been all night. I showed him where I had been sleeping and told him I was damn mad someone had slept in my hammock. He looked confused and said, " No one slept in your bed. I hung my hammock next to yours and I should know. " Then I looked and realized that I had walked into the wrong track. Joe smiled and said, "Lets go get something to eat". My mind filled with the thought of the normal breakfast we had when we were at this place. It was usually cold eggs and green ham. I threw up.

One week before Christmas, 1992. We were at a Christmas play with some friends. The play consisted of 7 actors that represented 7 countries. Each person told stories of Christmas in their countries, and they were very good. I was enjoying the play and the fun. One of the actors told a story about World War one. On one side, in the trenches were the British and in the trenches on the other side were the Germans. It was Christmas eve and the fighting had stopped. From the German side came the sound of one man singing. He was singing a Christmas song in German. A song that the British new. When he finished the song, three British soldiers began singing "Oh holy night". When they finished the entire German side started singing Silent Night. And so it went with each side taking turns singing songs.

This of course got the officers very upset on both sides and they began yelling at their troops to stop singing. Then, One of the German soldiers climbed out of his trench and began walking toward the British side. A British soldier climbed out of his trench and walked toward the Germans. Then all of the troops climbed out of their trenches and met in the middle of the battle field. They shook hands and wished each other merry Christmas. This made the officers extremely upset.

The man telling the story said that if, somehow the soldiers could have held on to the peace that happened that night thousands of men would not have died as the war continued.

One of the survivors of that night, a British soldier, later told of the wonder of that Christmas and the sadness he had felt, when he and the British soldiers had finally overrun the German trenches and he had found the bodies of two of the Germans he had met on Christmas.

I sat in that theater and I remembered what it had felt like, 24 years ago. When the officers told us stories about the enemy. Stories that may or may not have been true. What it had felt like to be waiting for war and how our leaders wanted us to hate our enemies. I remembered Joe, my best friend that Christmas. Joe was killed January 31, 1969.

I sat in that theater and a tear ran down my cheek. I wiped it away because I was embarrassed. A second tear followed the first.

CHAPTER 12

FEAR

"Oh yeah, Your Chicken." Never again would these words bother me. I was 14 and one of the older kids had said I was chicken. The reason was that I didn't want to climb a tree and tie a rope to a branch. We had many swings. Usually they were long ropes that hung from tall trees. We built them on tall cliffs. We were always looking for the perfect tree from the perfect cliff. This tree had held many swings.

Each summer we would hang a new rope. This summer I was getting the honor of hanging the new rope. I grabbed the heavy rope and started up the tree. Long before, someone had nailed two by fours to the tree to stand upon. It made a neat ladder out of the tree. Without these steps, carrying the heavy rope up the tree would be impossible. The rope was about an inch thick. The swing would be over 30 feet long.

After much sweating, I reached the thick branch where the swing would hang. I put both arms around the tree and passed the rope over the branch. I was bear hugging the tree. The limb was on the lower side of the tree so I had to tie the knot without seeing it. I made the first tie and leaned forward to make the next tie.

To do so I had to push with my feet and lean forward. The two by four step split in half and fell to the ground. As it broke, I felt myself slide around the tree. I was no longer on the top side of the tree. I was holding on for dear life. I watched the wood step fall the 35 feet to the ground. I felt dizzy and saw in my mind my body hit the ground. I was scared, very scared.

The kids on the ground started to give me advice. "let go of the tree and slide down the rope." "Just drop and we will call an ambulance. Don't worry it's not too far, you'll be OK."

I just sat there hugging the tree. After about half an hour in their time and a life time in my time, they left. My younger brother, Dave, stayed to see what would happen. As they left, they gave instructions to Dave to call them right away if I fell out of the tree. This made them laugh.

By this time, I was trying to figure out how to land so I would do the least damage to myself. Should I try to land butt first or land on my feet?

Dave yelled that I should move back real, real slowly. If I had been doing this all along, I would be down by now. Just move real slowly. I took his advice. Less then an inch at a time, I slid back. I wanted to get back on top of the tree. It was very, very hard because I was so scared that I had trouble making my body move. As I backed up, my leg slid over the nails left by the step. It hurt. I didn't care. I figured that if I fell out of the tree, that scratch would be the last things the medics would look at. Slowly, slowly I inched back. Finally the nails were at my thighs. This was not good. Any further and I would be in big trouble. I inched back until the nails were at my crotch. By now I was on top of the tree but still too scared to stop bear hugging. I lifted myself over the nails and felt my feet touch the first step. I was afraid it would rip loose like the other one. I was getting tired. I had no idea how long I had been in the tree. Slowly I moved from step to step. Finally, I was almost to the ground and I jumped free of the tree and landed flat footed. My knees collapsed and I hit the ground. I looked around. Dave had gotten bored when I reached the first step and had left.

I wasn't really afraid of heights after that. I was afraid of getting in that situation again and being that scared. Fear is not fun. In the army, I wanted to be a paratrooper. I had joined the army to be the best and to seek adventure. At least that is what I told myself when I signed up to be as paratrooper.

Paratrooper school was for the elite. Anyone could sign up but not everyone was accepted. I passed many tests to get to paratrooper school. I was proud when I was accepted.

The first week was basic paratrooper training and a lot of exercise. We ran miles, then did chin ups just to be allowed to go into the exercise area. We were not allowed to walk for the three week training period. We ran or we stood. No sitting, no walking. If we were ever caught walking, even on our time off, we had to do pushups.

The first scary thing I had to overcome was the practice for landing. We were tied to a rope and hung from a pulley about 10 feet in the air. The trainer would get us swinging, then he would release the rope. We had to land and roll. I didn't like the drop but I was quite good at landing and rolling.

The most difficult problem to overcome was jumping out of the "mock door". This taught us to jump out of a plane. They tied a rope to a parachute harness and made us jump out of a tower 35 feet high. We would drop about 10 feet and then the rope would get tight. The rope was tied to a pulley attached to a long rope that looked like a long clothes line. We would bounce down the clothes line to be caught at the end. We had to keep good form throughout the jump. We started with about 100 people. we had to make 8 perfect Jumps. Finally there were only three of us. I finally passed just in time to make the jump from the 250 foot tower. I wasn't looking forward to this.

We started early in the morning. No exercise period this day. The towers allowed three people to jump at once. We would be hoisted up 250 feet and then the chute was released. We would float down and make a perfect landing. Piece of Cake!

It was a very hot summer day in Georgia. The army kept the ground plowed at the bottom of the towers so our landings would be softer. The trainers held a competition at the base of the tower. The team that had their paratrooper hooked up first got to have a drink of water. My team had not been very fast and I was getting very thirsty.

They called the person two numbers ahead of me. Good, I'll get this over soon. I was rushing to get the parachute hooked up. I needed that drink. Then I remembered, that guy had never showed up. No one had that number. Then they called the next number. Without thinking, I yelled that he had dropped out last week. Then they called my number.

Hurry, hurry everyone yelled. They wanted water too. I grabbed the harness and threw my legs into the straps. The leader signaled that I was ready. I wasn't ready. The straps between my legs were not hooked up. I was just holding on by my arms. I quickly started putting the straps together and fastening them. At the same time I felt myself being lifted. Finally I was completed, Just as I reached the top of the tower! I was 250 feet in the air and MY FEET WERE DANGLING!!!

I suddenly realized I didn't really want to be a paratrooper! Next to my head, on the tower, was a speaker. He told me to release the safety strap. I didn't like that idea. There usually is a reason they give names to things. I wanted all the safety I could get. The man in the speaker yelled at me to hurry up. Good manners are not part of the army way.

After I released the strap he said, "let it drop". God, it fell a long, long time! Then he said, "Are you ready". I answered, "NO!" And they released me! At first, I thought they made a mistake. I wasn't dropping. Then I saw the shadow of the parachute on the ground getting larger. I was just starting to like it when they yelled at me to prepare to land.

I did a perfect landing (the army calls this a PLF. I think this means parachute landing fall.). How well I landed didn't matter. I landed in about 4 feet of dust. I completely went under the dirt and felt like I landed in water. Before I could do anything but choke, hands grabbed me and pulled off the harness. If they hurried, they could get a drink of water.

Covered with dirt and sweat, I walked toward the drinking fountain. "Where are you going, TRAINEE?" came the voice of my drill sergeant. "For a drink of water", I answered. "You didn't earn it yet. Give me ten pushups then get back out there in the sun with the rest of the TRAINEES."

I really didn't like the army nor the sergeant. The next day they gathered us for a formation. We sat on bleachers and they hoisted up various types of unmanned parachutes. We saw the types used in WWII and modern types. Then they showed us our parachute. It settled easily to the earth. Next, they showed us various problems that happen to parachutes. The first was a streamer. It was a chute that didn't open. Not a pretty sight. Then they showed us a May West. This happened when a person did not leave the door of the plane correctly. The paratrooper would actually go through his lines. Some of the lines would then run over the top of the chute. This decreased the actual area of the parachute and looked like a huge bra. Thus the name, May West.

Then they showed us our safety chute. This was a smaller chute that we carried on our front. They raised it up the tower and it didn't open. They tried again. On the third try the chute came streaming down and the man next to me jumped to his feet and yelled," I want to quit!

My first jump. I wasn't real scared although I had gone to church the evening before. It was a hot Georgia summer day. I wore glasses so they taped the glasses to my head. The tape did not stick well to the sweat on my face. I was the first one on the plane and therefore I would be the last to leave. The plane shook and stumbled. I wanted to jump from this old plane. I wasn't sure it would land. After we were in the air for a while, the sergeant said, "Stand up". There were no windows so I couldn't see where we were. "Hook up", was the next order. I hooked my strap to the cable above my head. The man next to me sat back down. Oh, well, I understood how he felt. The line of men began to start moving. It was amazing, The line just got shorter. One more man ahead of me. Then I turned and faced to open sky. Gulp!

The sergeant hit me on the butt and yelled over the engine noise, Jump". I wanted to turn to him and say "what?" But he hit me again and I jumped.

I expected to feel like I was falling. This was not the feeling at all. I immediately was hit by the prop blast and felt myself go horizontal. I saw the tape leave the side of my head. The strap at the back of my glasses held and they stayed on. Many times I had watched the leaves fall out of a tree on an autumn day. The wind would catch the leaf and whirl it around. That is exactly how I felt.

Then the wind stopped and I was hanging from my parachute, I looked up and it was in perfect shape. "Thank you, God".

I looked around and I was surrounded by my friends. We all were whooping and calling to each other. I was having a great time when I realized the ground was almost here. After being so high, 20 feet seemed like nothing. I prepared myself and waited. Smack, the ground hit me. I did a perfect PLF and started to gather my parachute. A man on the ground was yelling through a bullhorn at the soldiers that were just leaving the next plane. He was yelling "pull your reserve, pull your reserve". I looked up to see why and in the middle of the string of parachutes, one man was falling faster then the rest. He had a May West. Suddenly all the paratroopers released their safety chutes. Everyone except the May West. He was now close enough for me to see the safety chute wrapped around his feet. The man on the ground suddenly changed, "Pull your reserve" to "Prepare to land, prepare to land". An ambulance took off from behind me and arrived just as the man landed. The ambulance men ran to the unfortunate paratrooper. Then the rest of the soldiers landed all around them. One of the paratroopers landed on top of the ambulance, did a perfect PLF, rolled off the roof and hit the ground and did another PLF.

The next day before we made our next jump (we made five, total) we were told that we had to keep good form when we left the door or we would also have a May West. The poor man had broken both his ankles. "He had not made a PLF but landed flat footed. We all felt very, very sorry for him. Not because he hurt himself but because he had gone the three weeks of training and could never be a paratrooper. At that time in my life nothing was as important as being a paratrooper.

Eight months later, I was in Vietnam. Our job was to guard a company of national guardsmen that were working on the roads.

They had to clear 100 feet on each side of the road. This was going to stop ambushes on the road. They did not carry weapons. That was our job. Most of the time this was not much fun. We sat on some dusty road and watched them work. I have always hated watching someone work. I want to help.

Their lieutenant had to move up and down the area to see how his troops were doing. At night we would sit around with the guardsman and got to be good friends. Their lieutenant didn't like driving around with us, so he ordered a jeep. The jeep had a machine gun mounted on a pole. Just like rat patrol.

When the jeep arrived, One of my friends, a black man, was picked to ride around on the machine gun. He was very happy because he had the best job. He just road around looking important. We tried and tried to convince these people that the enemy was always around and watching us. They just didn't believe us. Every day that we didn't see combat, they were more right and we were more wrong. I tried very hard to tell this man what a dirty trick they had played on him. That piece of machinery was good for television but it was worthless in this war.

In combat, Americans do not focus. They fire off thousands of rounds of machine gun fire hoping to hit something. We had an old sergeant that tried to make us focus. Pick a target, fire at least one burst and the pick another target. Do not spray the countryside because it doesn't do any good.

The Vietcong were very disciplined. They bracketed a target. Two gunners picked a person and fired at him until they were ready to move to another target. In more then one battle, someone would yell, get them off me, while the rest of the people would not be receiving any rounds. Of course, the machine gun on a jeep would be the first target. And where would he hide? One day we heard an explosion and rounds being fired between the two areas we were guarding. We took off toward the sound. As we came around the corner, the black man walked out of the brush. The front of his pants were all wet. He tried to talk but his teeth wouldn't stop chattering.

Finally, he told us what had happened. Three Vietnamese solders had been standing on the side of the road. They had waved at the jeep as it approached. When it was a few feet away the three men had jumped off the road. Before he could figure out why they had jumped, there was an explosion. They had thrown a grenade. The grenade was supposed to go off under the jeep but instead had rolled too far and when it did explode it threw the gunner off the jeep and into the bushes. The enemy, dressed as Vietnamese, came up shooting. All three had hit the driver. When he was hit it threw him in front of the lieutenant and protected him. The driver had stayed conscious long enough to get the jeep out of range of the bullets that were sprayed after them.

The Vietcong knew that there was a soldier in the bushes. They did not know he was unarmed. They were looking for him and were within a few feet of him. He wet his pants. The odor was so strong, he was sure they could smell him. He almost passed out because he was not breathing.

Then he heard our track coming down the road and he saw the three enemy soldiers run into the brush. We fired a few rounds of 50 caliber into the brush and took him to the medic. He was physically OK but emotionally he was a wreck. The driver died before the dust off chopper arrived.

In combat, I did not feel fear. I was not scared, ever. In combat, the adrenalin rushes through your body at such a rate that you are a superman. I seemed to be able to see everything. I was aware of all the action behind me. I could see in the dark and my mind was very, very clear. I was incredibly strong. And I never missed a target. My thoughts came very, very fast and I made immediate decisions that I knew were correct. Time seemed to stand still although, later, I was always amazed how many hours went by.

The next day I was hit with extreme exhaustion. All muscles ached and my mind was fuzzy. I just wanted to sit. If adrenalin is a drug, there is an adrenalin hang over.

Fear, in Vietnam, came with peace. After the battle, I felt confused. I would be trying to cope with my friends dying. Also, I felt good that I was alive. After a few days, I would feel at peace. I knew that we had hurt the enemy and it would be awhile before he came back. This period of peace was wonderful. I wrote a lot of letters to my family during this time about how good life was.

As time went on, I knew that the enemy was setting up an ambush. They would set up, so that anyone entering the ambush would be in trouble. Then they would perfect the ambush. I was told that they would do this for years, just waiting for someone to walk into their trap. Each day the trap got better and better. Each day, I and my friends became more uncomfortable. Finally, combat would come and the cycle would start over. To help matters, the army would send us to different areas, just to force us to fall into a new trap. Also, we had special places where we always drove and special places where we always spent the night. This helped the enemy set up traps. It was a terrible cycle set up by the army and the enemy.

After I got out of the army, I got married. I married a good woman that loved me very much. We have two children. I was 22 and she was 18 on our wedding day. A few weeks ago I found our wedding pictures. I was shocked at how beautiful she was on our wedding day. We had only dated 5 months when we were married.

As I looked at our pictures, I noticed how tightly she held my hand as we said our vows. The photographer caught us just right. She was uncertain of what was ahead, but I knew she loved and trusted me. We would have a good life together.

Every few months, I would start feeling uncomfortable. We would have a fight. At first she tried very hard to solve what was making me so mad. I would wake up in the morning, after having nightmares about Vietnam and open my drawer to get dressed for work. If I didn't have any socks or they were not put away just right, I would scream at her. Then after the fight, I would start to feel better. At least for awhile. She tried and tried to make me happy. For 17 years she put up with my raging and my anger.

One day I told her I couldn't take it any longer. She had changed and she was suffocating me. I wanted out.

Five years ago I got a divorce. I left very angry and today, I can't remember why. I started a new life with a woman that will not put up with my anger nor my rage. When I would start a fight, she would tell me to get out.

Later I would return and everything would be great. After our third or fourth breakup, she told me to see a Vietnam vet counselor or get out forever.

After one and a half years of counseling, I discovered that the combat and peace cycle I learned in the army had become part of my life. In my job and in my relationships, as soon as everything becomes comfortable, I begin to get scared. If the peace lasts very long, I will do anything to cause a fight.

Good news. Now that I know the answer, the problem is solved. I wish that were true. Now it is worse. I see the signs of peace creeping into my life. Now I understand what is coming and I prepare for them. I try not to fight. Remember the terrible enemy? The longer there is peace, the worse the combat. The longer I can keep peace, the more troubled I become. Suddenly I find myself asking, "Why am I putting up with this? That is a clear sign to me that something is going to happen. Then the terrible nightmares return. Some about Vietnam, some just violent. I have not yet gone over six months with out a fight.

So, during one of the peace times, I sat her down and told her about this. In my life I have never been so outraged as when she told me, during an argument, that I was just going through a "Vietnam thing." I want peace in my life. I want to share a home with someone. I asked her to marry me, and she said, "No". When I asked why she said that she didn't want divorce. Smart woman.

So, in one of my group sessions, I asked the group what I should do. How do I control these feelings. They answered that there is only one way. To do as nine of the eleven of them were doing, Live alone. One of them has a good marriage and the other is getting a divorce. There are statistics on how many people died in the war. I wonder how many people were hurt?

CHAPTER 13

ANIMALS IN WAR

I was sitting in an auditorium at Washington State University. The class held over a thousand students and was full. We had taken a practice test and the tests were being handed back. The class had desks in rows in a theatre type setting. The floor sloped down to the stage where the professor was standing, arms crossed.

In college I could take elective courses. These courses were designed to give me a rounded education. One such course was Environmental Science. I was interested in the environment. I found that taking courses that interested me was to my benefit. I would get better grades and this helped my grade point. I had gotten on the honor role and planned to stay there.

The professor called these tests "quizzies". He thought he was being cute. His "quizzies" were designed to give us an awareness of where we needed to improve.

As the "quizzies" were being returned, I was a bit uncomfortable. The "quizzie" had been essay type and, until the paper was back, I wouldn't know what my grade. High grades were very important to me at that time. As each person was handed their "quizzie", they became very silent. Not a good sign. The class of over a thousand is, by nature, somewhat loud. As the names were being called out, it was getting strangely quiet. Finally they called out, " Manthey" and I raised my hand. I was not expecting the score I received. On the top of the paper was a ten. That meant 10 out of 100. Under it was a note, "you obviously do not know the material". I looked over the persons shoulder in front of me and saw a two. The person next to her had a five. The class was getting very quiet. I was thinking that perhaps taking this class was not a good idea. Instead of helping my grade point. I would be hurting it.

A name was called and a girl two rows ahead of me raised her hand. By now the only sound in the room was the names being called. When she got her paper she sat down and in a very clear voice said, "If this is what his quizzies are like, I can't wait to see his testes". There was a short silence and the entire class, including the professor burst out laughing.

I stayed in the class and eventually got a B. After the class stopped laughing, the professor started his lecture. In college, no one cares how you did on a test. They just keep on lecturing. His lecture that day was about endangered species and why they were becoming extinct. The slide projector came on and a picture of a Condor was shown. These are very ugly birds. I raised my hand and asked if these were only found in California. I was surprised to find that they were. I had seen an ugly big bird that looked just like a condor in Vietnam.

It was during the very dry part of the year. We had sat for many days. Everyone was board. For fun some of the G.I.'s would roll grenades down a hill and bet on whose would roll the farthest. Then they would measure the hole in the ground and the winner would get the money. This was stopped when and innocent bystander was wounded.

Suddenly, a rifle round was fired. I looked at the center of our perimeter and three men were looking up. One had just fired. I looked up and saw a bird slowly flying overhead. It looked about the size of a sea gull. The bird started to fall. It had taken an incredibly long time for the bullet to get to the bird. As it fell, I realized it was a huge bird that had been very high in the air. When it finally hit the ground, we measured a wingspan of about 10 feet. It was a very ugly bird.

I have heard it said many times that the ecology of a country suffers during a war. It is not the war, but the large number of men walking around with weapons, that destroy the environment. I have often thought that a course should be given to all soldiers going to war on the repercussions of destroying the environment.

One day we were out driving around in our tracks when the left machine gunner on the left track opened fire. We turned our vehicles to face the enemy. All machine guns were loaded and ready. He had seen a family of wild pigs. A huge sow and four babies. He killed all of them.

Another day one of the gunners opened fire and killed a giant lizard. These look like Iguanas and are about 12 feet long.

During the rainy season we were driving through the rice paddies when a Python suddenly raised his head to the level of a man on a track. The python measured over 30 feet long. The man sitting on the track next to the snake opened fire with his M-16. We gave the dead snake to the Vietnamese. They cut it into slices and cooked it on a stick. I tasted it. You have heard people say that snake tastes like chicken? This one tasted like very salty shoe leather.

Anyone that has been in Vietnam will tell you stories about rats. They are huge and are everywhere. One day I determined I would sleep in a bunker. These are made of sand bags and have walls about 6 feet thick. I felt I would be safer then in a track. A recoilless rifle would easily go through the side of a track but would explode in the walls of a bunker. The problem with a bunker was that they flood when it rains. So I took a stretcher and stuck it in the sandbags near the door. No one else would sleep in the bunker. I felt fairly secure as I crawled onto the stretcher that first night. In the middle of the night I awoke with something small sleeping on me right knee. I could feel tinny hands digging into my knee cap. It was very dark. I sat up and a rat jumped off my knee onto my chest. I screamed and he made another leap, right on my head!

The next night I armed myself with insect repellent. This is very powerful stuff. I would have slept with my rifle but I was afraid I would shoot myself in the dark. Suddenly I awoke. He was back. I didn't Move. I felt for the insect spray in the dark. Rapidly I aimed the spray at my knee. The rat jumped just in time. He landed on my chest. I screamed and he landed on my head. I could feel the tiny feet digging into my scalp. I swung the bug spray around and fired. By then the rat was gone and I sprayed myself right in the face. I was yelling so my mouth was open!

After that night, I let the rat sleep on my knee. It gave me the creeps but it also gave me the creeps to think of him on my head. I tried using traps and poison but the rat was too smart. After about two weeks of sleeping with a rat I began to get used to it. When morning came, he was always gone.

One morning I was having my favorite breakfast; Canned ham and canned pineapple from the "C" rations. I looked up and two cats were walking down the road, one on each side of the road. It is very rare to see cats. The Vietnamese eat them so they stay far away from people. These two looked like they were on patrol. They walked into my bunker. They were in the bunker for about 15 minutes. They came out, looked at us and walked into the bunker on the other side of the road. When they were finished, they went down the road.

That night the rat was gone. I slept in the bunker for the next two months and the rats never returned. Tigers are as scary as rats. We had been sitting in our perimeter for about three months. Each day between two o'clock and four o'clock, we received enemy rounds. We did not bathe in the river around this time because we had to return fire with our machine guns, and this was not comfortable when we were wet. Finally, our captain had enough and he called in the Air Force. They flew over us and dropped napalm. This completely cleaned off the hill including all vegetation. We were convinced the Viet Cong were in the hillside in caves and bunkers. We guessed that he had been cooked.

About a week after the napalm was dropped, I was asleep. My best friend, George Alvin Tyler the Third, woke me up. I was angry. I didn't have guard for three more hours. George was a black man. In the dark all I could see was his eyes and they were huge. Steve, Wake up. There is a tiger out there. "Right" I answered.

I went on guard with George. He guessed the Tiger had been attracted to the cooked Viet Cong and now they were coming down to visit us. I laughed at George. I told him he must have fallen asleep on guard and dreamed about a tiger. Anyway, I was up and I would sit on guard for him. He could get some sleep.

He had been gone about an hour. I was sitting on the top of the track with my M-16 on my lap. I could see fairly good because there was a half moon. Suddenly I heard a very loud Roar. The hair stood up on the back of my neck. I was sure the tiger was only a few feet away. I opened fire with the M-16 and when that was empty I began firing the M-60 machine gun. I fired 150 rounds into the night before I was through. Then I sat down and waited for morning. I kept trying to figure out how to get my new tiger skin home. When morning arrived I woke everyone up. I told them about the tiger and said we had to go look for the body. I didn't want to do it alone. What if he was wounded?

We began the search. The ground was full of bullet holes where I had fired. We looked for the tiger. Then we looked for the blood. Then we gave up. Actually they gave up. Only George believed I had heard a tiger. We never heard the tiger again.

The Vietnamese have small animals. Their cows are the size of our dogs. Their dogs look like full grown German shepherds but are the size of a cocker spaniel. All their farm animals are small, except the water buffalo. I thought the Vietnamese, being a small people, had down sized all their animals so they could control them. This made me wonder at the size of the prehistoric water buffalo. He must have been huge.

One day, I was driving point track. We were far into the paddies. I spotted a water buffalo with a young calf, probably about a week old. For fun, I drove my vehicle between the calf and its mother. I was having a great time playing cowboy. Suddenly the 15 ton track jerked to my left. At the same time I heard a bell ring. My track commander came over the intercom and said the water buffalo had charged and hit us on the side. I should leave the calf alone before someone was hurt. About that time, we were knocked sideways by the charging buffalo. The huge horns ramming against the side of our vehicle sounded like a bell ringing. Later I was in a collision between two tracks. The water buffalo hit us harder then the track.

I was always trying to shoot birds. I could usually hit them in the air with my rifle. One day we picked up a choe hoy. Choe Hoy is

Vietnamese for surrender. This was a Viet Cong that came over to our side. They did this when they were hungry. To prove they were on our side they would show us an enemy camp. We drove to the site of the enemy. His camp was on the side of a hill. We turned our five tracks toward the hill and opened fire with our machine guns. To get the best shooting we would aim the 50 caliper and one M-60 toward the hill. This meant one gun could not fire in the direction of the enemy. My gun was not being fired. I looked at the hill and saw a white bird. Goody, a target. I began firing at the bird. It was flying up and down and dodging the bullets. Finally, after about a hundred rounds, I hit it. By then the shooting stopped and a couple of my friends came over. They had seen me shooting at the bird and were impressed that I could shoot one out of the air, that far away.

As was standard procedure, we carried Vietnamese soldiers to do the dirty work. They were very reluctant to climb the hill. They climbed straight up the hill to the enemy camp. Suddenly, on the hill, I heard a "Womph". This was a B-40 rocket being fired. A B-40 is like a Bazooka. It was being fired at the right flank of the enemy's location, right where I had been shooting at the bird!

As it turned out, The choe hoy had maid a mistake. The enemy camp was to the right. When the Vietnamese climbed the hill they found blood and pieces of flesh and even body parts. Apparently, while I was happily shooting at the bird, my bullets were finding people. Two Vietnamese were killed climbing the hill by the B-40 round. I have often thought of the poor Viet Cong seeing the bullets come in and laughing at the stupid Americans for firing in the wrong location. Then the bullets coming at a bird. Maybe one of them shot the bird to make me stop.

Vietnam is a beautiful country. The ocean is warm and clear and their beaches are over a mile inland with clear white sand. The weather is always warm and the hills are full of wild animals.

I used to think it would be a perfect place for a hunting lodge. After I had been back from Vietnam about two years, I had the strangest feeling. It was a feeling like homesickness. Many Americans loved that country. The country and the people are like the Vietnamese jungle. It is not like a jungle where Tarzan lived. It is just a very dense brush. They do not have tall trees. We were told to guard the engineers to clear a road. This road was used by the enemy and ourselves. We were always getting hit on this road. If we could see 100 yards on each side of the road, we could see the enemy and get a bigger body count. So, they cleared 100 yards on each side of the road, right down to the dirt. Before we were a mile up the road, The brush was growing back. By two miles, It was as dense as before we started clearing. By three miles, It was denser then before. We did not kill the jungle and we did not stop the people. Not all things in the war were bad. I remember many good times. They were the scariest of times with the best of friends.

CHAPTER 14

UNEXPECTED

"The greatest example we have of death is birth", I explained. It was a beautiful August evening and I was on a sailboat with an old friend. We were sailing into the sun. The air was salty fresh and the sun was warm. There were just two of us on the boat. "When a person is in the womb, they are safe. Birth is very painful for the baby. But, after it is over, the baby forgets the pain and is exposed to a glorious world of light and beauty. That is the same as death. That is what we can expect."

My friend smiled and said, "You are right but you are wrong." We have been friends for over 18 years. We had met in college. Years before, he had been at a tavern with a friend. They had decided to set a new record. They would travel on a motor cycle to a nearby town and return within a short time.

They were traveling over 100 miles an hour most of the trip. At one corner, they lost control of the bike. As the bike started to lay down, my friend noted their speed was over 90 miles an hour. The driver was pulled under the bike and was twisted and broken. His bones were sticking out of his legs. My friend left the bike and crashed into a fence. A barbed wire fence. He traveled down the fence at a great speed, cutting his throat on the barbs. After he had laid on the fence for awhile, he felt himself being lifted up. As he looked down, he saw his body. He could see the blood flowing from his neck wound and he could even see the misquotes and gnats buzzing around his body. He could also see his friend. His friend was dragging his broken body back to the road to get help.

He was lifted up, up to a bright light. It was a light but not a light. He also saw people that he knew. They were very happy for him and wanted him to stay. He convinced the light that he wanted to go back, that he had something he had to do.

By now the police had arrived at the accident. They looked at my friend, draped over the fence and they noted that there wasn't any blood coming from his wounds. They declared him dead at the scene of the accident. He was covered and carried to the ambulance.

Suddenly my friend sat up and declared that he was back! The ambulance medic was working on the other man's broken legs and said, "You can't be back, you are dead!"

My friend lived. He has a scar under his chin, down to his adams apple and back to his right ear. "Yes", he said, "Death is like birth. But it is not an explosion of your senses. It is not like coming from the womb into the world. It is more like going back to the womb. close your eyes and feel your body. Feel the pressure of gravity pressing against your chest as you breath. Feel each muscle and each pain. At death, all this is gone. You are so used to feeling your body that it is hard to describe when it is gone." Then he said, "You are afraid to die, aren't you".

"Yes, I am very afraid of death. I have seen so much of it and it is always so awful. I have spent much to much time worrying about death. But, if it is like you say, then I will probably laugh when it finally comes to me." "It is more likely that you will feel relieved and wonder why you were so afraid," He answered.

We sailed on into the evening. I felt a growing closeness to my friend and was completely enjoying the sail. Also, I felt an old tension lifted.

If death is so wonderful for the person that died, why is it so difficult for those left behind? When my grandfather died, I felt that it was a good thing. He had lived a good life and he had been in so much pain at the end. Also we had years of preparing for it.

Death, when you are not prepared, is a terrible shock.

Each day, in the last few months of Vietnam, I would look at the men around me and wonder if today was his last day on earth. If he dies tonight, how will I feel. And, if I die tonight, how will he feel? It meant that I began to limit my number of friends and I began to enjoy everything I did with my old friends. But, even in Vietnam, there are surprises.

He was the funniest man I have ever met. He was in another platoon. He was black. His head was too small. His hips were too big. His face was covered with a very wide nose. His eyes were too close. He even walked like Groucho Marks.

It was as if God wanted to create the perfect comedian. His sense of humor was wonderful. His timing was great. And he was very creative. To sit with him for over an hour was too long. I would always leave with my cheeks aching. He was funny.

We were having a normal Vietnam day. We drove our armored personal carriers (tracks) to a place. Sat on guard all day and then returned. Suddenly, we got a call that a track had hit a "LIMA MIKE". This is code for land mine. It was the second platoon and we were the closest track. I was the driver. I raced down the one lane dirt road. We were about a mile away. We were always afraid that Charlie would attack after a land mine had blown up a track. It didn't happen to us-ever. Thank God! After a mine explodes, there are wounded everywhere. The track can't move and you are a sitting duck. Most of all, everyone is stunned.

We rounded the corner to find the track with the back gate open and people talking to a wounded soldier on the ground. No one was on the machine gun! The second platoon rarely were hit. I was in the first platoon and we were always under fire. I couldn't believe it. I turned the track towards the East. All of our battles on this road had been from the East. I heard the Track Commander pull a round into the chamber of the 50 caliber machine gun. Before I was stopped, he was firing into the brush. When I looked at the men on the ground, I could see fear in their eyes. By now both m-60 side machine guns had joined the 50 caliber and were firing into the brush. To have trauma and then to hear machine gun fire was difficult. Also, they suddenly realized how stupid they had been. Stupid! In Vietnam that got you a purple heart or death.

Unless we started receiving fire I, as the driver would not be needed. I had to see if the soldier on the ground needed help. I jumped off the track and ran to the wounded person. Have you called for a dust off? No! Is there a secure place nearby where we can land a chopper and get this man out of here? No one answered. Just dumb looks. I ran back to my track just as another vehicle rounded the corner. Good, I thought, its track ONE-3.

This was the track with the first platoons medic. A good man. He would wake up these people. I jumped onto my track and told the track commander no one has called for a dust-off. He had already located a clearing on the map nearby and began to call for the dust-off. I ran to the back of our track and grabbed the stretcher. I then ran back to the wounded man. Oh shit! It's Lamont. Lamont was the name of the comedian from the second platoon. His leg was broken and he was sitting up. He was cracking jokes! As I arrived with the stretcher, he cracked a joke and everyone laughed. The medic turned to me and said we had to get him to the dust-off, but not to worry. He wasn't hurt bad. Just a broken leg. We loaded him up and put him in our track. Then we met the dust-off. The last time I saw Lamont, he was being loaded on to the chopper. As the medics loaded him on, he said something and they all laughed.

Two days later we got the word. Lamont had died. At first, we were told that he had died of shock. Then we were told that he had died because someone made a mistake at the hospital. I refused to believe it. I refused to believe that such a gifted man had died. NO! IT CAN NOT BE TRUE!

Mourning for someone that has died is a process. There are stages that everyone goes through. Some people take a long time with each stage. Some move through it very fast. At first you can't believe it. Then you are angry. Then you are sad. Most people don't cry until a later stage. Finally, eventually, you accept it and realize that the person is gone forever and that you were lucky to have had time with them. We couldn't go through this process in Vietnam. I didn't mourn for Lamont. I didn't say goodbye. I still haven't dealt with his death.

Years later I was in college. It was summer and I was enjoying life. My second child had been born that June and I had an easy class schedule. I was sitting on the deck of our apartment looking at the Eastern Washington Summer. My wife had been shopping and had just arrived home. She was carrying the mail. "Look Honey, a letter from your Mom."

This was exciting. With two kids, we were living on the money I received from the GI bill. This was $316.00 per month. There wasn't enough money left to pay for a phone. I hadn't heard from my Mom for awhile. Dear Steve, I'm afraid I have some bad news for you. Sharon was killed in a car accident. I know how close you two were. Her funeral is on Saturday. (it was Monday. I had missed her funeral).

I have no idea what the rest of the letter said.

I got up from my chair and handed the letter to my wife. I smiled and told her I needed to go for a walk. I was numb. I didn't feel sad or upset. I didn't feel anything. I walked out of the apartment into the sun. A sun that Sharon would never again feel. I put one foot in front of the other and walked. I walked out into the wheat fields and I walked down dirt roads. And I remembered all the great times I had with my cousin. Yes, I had been close to her closer than anyone really knew. She was my first love and had been my friend.

I found myself walking into the game room in the apartment complex. It was large, dark room without any windows. The air conditioning was on and I felt a chill as I entered. I grabbed a pool cue and set up the balls. I aimed the cue ball at the group of balls and I fired. I was surprised at the force I put behind the hit. When the balls stopped moving, I laid my head on the table and I cried.

I didn't talk to anyone about Sharon. I wouldn't talk to my wife and I didn't ever ask my family what happened. I couldn't. Once in awhile, I will see someone that reminds me of Sharon. Once in awhile, I will hear a song that she and I used to like. Once in awhile I will miss her.

Thanksgiving, 1975. I was out of college and had my first job as a designer. We were having Thanksgiving at my mother's house.

My brother and two sisters, with their families, were there. Just as we sat down to eat, there was a knock on the door.

It was my father. He was just stopping by. For years I had been very angry at my Dad for divorcing us and my mother. I had just started to get to know him and things were a lot better. It was a little uncomfortable having him there. Dad said that he was going snowshoeing the next day and asked if I wanted to join him. No, I answered. Maybe next time. To bad, he was going with his boss from work, his son and my half brother. My father explained that he didn't like his boss. He said he was crude and overbearing.

He said he would let us get back to our dinner and he left.

The next night I was sitting at home and the phone rang. It was my fathers wife. She said there had been a accident on the mountain. My father had not yet gotten home but she thought he was all right. What she said and the tone of her voice did not match. Something was wrong, very wrong. Before I could ask any questions, she hung up.

I called the police. They didn't know anything. I called the forest service. They didn't know anything. Then the phone rang.

It was my Dad's best friend. He was at my Dad's house with Merrily, my Dad's wife. He said that the forest service and the police had called him and asked that he call me.

He said," Yes, there had been an accident." The man that was snowshoeing with my Dad, his son and my half brother, Scott had gotten out but my father was not with them." They were waiting for more news. He would call as soon as he heard anything.

I got off the phone and told my wife that everything was OK. Everyone had gotten out of the woods except my Dad. This was good news. My Dad knew the woods and never got lost. He would soon be out and everything would be fine.

Two hours later the phone rang. It was Paul, the man I had talked to earlier. Scott had just arrived home. The man that had been with Dad said that there had been an avalanche. Yes, my father was dead.

They had left him in the woods and would get him the next morning. The police asked if I could be there when they brought him out... to identify the body. I said I would meet Paul in the morning. I called my brother and told him everything. Dave wanted to go with me the next morning.

That day at the ranger station is very confused in my memory. We seemed to wait forever for them to bring my Dad's body out. During this time I had a chance to take the man aside that had been snowshoeing with Dad. I said, " I want to know everything, do not leave out any details. I was in Vietnam and I can handle it."

"We were snowshoeing along a side hill. We had seen two avalanches and had been thrilled at how powerful they were. The two boys were in the back. Your father was leading. There was a rumble and I looked up to see a wall of snow. It took your father and me down the hill. The two boys were buried up to their shoulders but stayed on the trail. I was buried under the snow but managed to get my hand in front of my face. This gave me an air space SO I could breath. I was only a foot under the snow but it was very heavy. It took a few minutes to dig out. As soon as I dug out, I called the boys and told them to stay put. They were just digging out. I found a ski pole and started poking around. I located your father and dug him out. He was about three feet under the snow. His snow shoes were still on. He was face down with the snowshoes over his back, his knees bent. He was not breathing. I rolled him over and his eyes were glazed over. I poked my fingers into his eyes and they didn't blink. I knew he was dead."

I asked, "Did you try mouth to mouth or CPR?" His face showed a look as if this was a very disgusting idea. "No", he answered. "Do you know how to give CPR?" I asked. " Yes, but it wouldn't have done any good, He was dead."

He must have seen the rage in my eyes. He turned and walked away and we never spoke again. It has been proven that people that drown in ice water can be revived after a very long time. One case, a young girl, was revived after almost a half hour. My father was in cold wet snow. If my father had been the one that survived, I know he would have saved his boss. If I had been there, my father would be alive today. I have had many years of first aid training. As I stood there on that day, I knew all of this.

The forest ranger walked up to me and introduced me to the coroner. He asked me my fathers address, age and his birthday. I could not answer any of his questions. I asked him to talk to my brother. Then he asked if I would like to see my Dad. I said, "yes". He led me down a hall and out a back door. Parked in the back was a red station wagon. I do not know if it was in a garage or outside. Everything around the car seemed black. He opened the passenger door and I sat in the seat. There was a blanket over the shape of a person. Sticking out of the end of the blanket were my dad's boots.

Funny things go through your mind. I looked at those boots and almost laughed. As I was growing up, my Dad would take my brother and me to get "hunting boots" at the Piggly Wiggly store. This was where you could get good cheep boots. By the time I got them broken in, they were worn out.

The man pulled away the blanket. My first reaction. There has been a mistake! I was used to seeing people that had been shot. I had assumed that part of death was the change of color from flesh color to gray. My father had his normal color.

Then I saw that he was not breathing. That familiar feeling of suffocating came over me. I looked at him for a little while longer and said, " Daddy, I will miss you."

I walked back down that long corridor and walked to my brother. The forest rangers were giving him my fathers things. I started to sob, deep uncontrolled sobs.

When we got back home, I assumed the leadership role. I tried to be tough. I tried to be numb and busy. I contacted all the relatives to tell them of my fathers death., I did a very poor job.

I tried to comfort Merrily. She just wanted to be left alone. She made me promise that I would take care of my fathers ashes and that he would never be buried in any place where she could go. I had already discussed cremation with my father and knew his wishes. Merrily was not thinking clearly and was in shock. I did what she wanted., I should not have listened to her.

My mother became very strange. She kept wanting to fight with me for control. She said that Merrily had been married to my Dad for 12 years and she had been married to him for 14 years so she should have a say in what was happening. She wanted a wake and she wanted my Dad buried in the church. She wanted a real funeral. My mothers presentation was not good but she was right. If I had to do it again, I would have listened to her. Instead, I fought with her. When she couldn't get her way with me, she attacked my wife. I went home to listen to all the terrible things my mother had said.

My sisters simply became shadows. The youngest had lived with my dad and now had to be moved to my mothers. I had to do this because my mother would not go to that "other woman's house". I felt very sorry for my sister. Not only had she lost her father but her entire life changed. In one day.

My other sister was there but not there. I recognized the pain she was feeling but could not deal with it. The second day I drove to the funeral home to pick up the remains of my father. I went alone. "Mr. Manthey, you are aware that it is against the law for you to do anything with these remains. ""No, I was not aware of that. I would like to take my fathers ashes up into the mountains and spread his ashes. This was what he asked me to do." "I'm sorry, but that is impossible. The law protects the environment and specifically states that a person must be buried in a cemetery I can, of course, give you His remains and you can transport them to another funeral home. You must sign this document and you must agree to deliver." I looked at him and signed the document. I kept the remains.

For years, I had dreams about someone knocking on the door and demanding my fathers ashes. I called all my family and asked if they wanted to go with me to bury Dad. Only Teresa the youngest, would go. We met that Saturday. All the mountain passes were closed. There was flooding and slides. The weather had turned bad and as I picked up Teresa, I felt like I could reach up and touch the heavy clouds.

Finally, after driving for many miles, we drove to a road that I knew. It was not very high in the mountains and it didn't have a great view. I was disappointed in the choice of locations, but I had to do something. Teresa and I walked up the hill. I carried a shovel to bury the box that carried my father's ashes.

At the top of the hill, I pulled out the Lutheran hymnal that I had borrowed from a church and started reading parts of the section on funerals. "Ye, though I walk through the shadow of death, I will fear no evil." This stuff started sounding like a bunch of crap.

I stepped up on top of an old stump and opened the box to throw the ashes down the hill. Below was a green hillside of oregon grape and above were many old growth trees. It seemed alright. I looked inside the box. Oh my God!" Teresa, the box is not full of ashes, its full of bones!"

I jumped off the stump and grabbed the shovel. I started digging. Teresa stopped me. "No, we are not going to bury our father like he is a dead cat or a small animal. Stand up and spread his ashes like you said you were going to do!" I looked at her and I looked at the box of bones. I stepped back on the stump and threw the bones down the hillside.

Then I closed my eyes and said, "Dad, I know you are in heaven. I know you are in a better place." Oh God, please take care of my father." As I said this I felt warm. Inside, I was not convinced my dad was in heaven. Teresa said, "Look!" I opened my eyes and the clouds had parted. The sun was shining through the trees and I was surrounded in warm sunlight. As suddenly as it happened, it was gone and the sky returned to the dark gray, low clouds.

For years I have tried to deal with my fathers death. It is hard to deal with the feeling that he did not have to die. When my friend told of his death and his return, an image played in my mind. I imagined being my father, there above the snow, watching the man dig him out. My father had terrible back pains most of his life. With the snow shoes where they were, he must have been in terrible pain when he died. There he was, with my grandfather and many of my Dad's best friends. I imagine he looked down and didn't want to come back.

When I returned with my sister, there was a letter from my Aunt, Sharons mother. She wrote a letter that I have kept in my heart. She said," You will never forget how you feel about your father. You will never stop feeling sad when you think of his death. But, eventually, you will be able to live with it. He has been gone 17 years. She was very right.

CHAPTER 15

BEST FRIENDS

What is a best friend? When does a friend turn become a best friend? How does this relationship start and when does it end? I don't know. 1969. Why? I do know that I have not had a best friend since...To answer this I will explore three people. My best friend ever, My last best friend and a best friend that I lost.

As I look over my life I see many friends that have come into my life. Many were close and many influenced me. The person I remember as the best, best friend, I didn't realize was even a friend until I started writing.

My father had a rule. It was a very important rule. All adults had to be labeled Mr. or Mrs. All relatives were called Uncle or Aunt. There were no exceptions. I remember calling Uncle Harold's first wife "Aunt Judy" and she was not real happy about it. I was 9 or 10 and she was 16.

Labels are important but they can get in the way. Now that I am 44 I still call all my uncles, "Uncle. . . ." It is funny to remember, but my father had us call him," Daddy." To me, this name took away respect. I guess he really wanted to be closer then we were with other adults. As a fifteen year old, I struggled with my identity and puberty. During the summer I was sent to the farm to work. Today, kids would want a good wage plus benefits and days off. I worked for my room and board and I worked hard. I loved it. These were the greatest days of my life. At that time two of my Uncles ran the farm. One worked as a logger. He worked on the farm part time. My cousin and I worked for Uncle Benny. My cousin, Jimmy was about 10. He worked very hard by modern standards but I thought he spent most of his time trying to avoid work.

So, in many ways, Uncle Benny and I ran the farm.

Each day would start at sunrise. He would get the barn ready for the Cows and I would bring them in. I would then change the sprinklers and we would meet back at the house for breakfast. Breakfast consisted of oat meal or Jimmy's pancakes. He had his own mix. We would drink coffee and eat pancakes. They were the size of the pan and about an inch thick. They were so heavy they would bend the pan cake flipper. No matter how much syrup was used, they absorbed all of it. Two pancakes and we were good until noon. During breakfast Uncle Benny and I would plan the days work.

We would talk about the most important chores to be completed and he always respected my advise. Slowly over the summer we became friends. He was 29 and I was 15. He was my Uncle.

Uncle Benny was a dreamer. He dreamed about a farm that would make him rich. He planned huge herds and was always trying to figure out how to get a higher yield. About half way through the summer he decided milking the cows three times a day would give more milk then milking them twice a day. We talked about this for days, always discussing the ramifications. Would they really give more milk? Would it shorten their milk giving life span? Would it cost more? What time of the day should we milk them? The discussion raged on. Finally we set up an experiment. We would milk them earlier in the morning. The second milking at lunch and the third time was later in the evening. We would closely monitor the amount of grain we gave them and we would check the amount of milk at the end of the month.

As I look back, we probably should have published our findings in a journal.

The results: Less milk per milking but a small increase over all. The increased income from the milk did not pay for the increased amount of grain that was fed to the cows while they were being milked. The other result was that we were working from four o'clock in the morning until ten o'clock at night. The two of us worked very efficiently together and I think Uncle Benny would not have wanted to keep doing this after I went back to school in the fall.

Milking the cows with Uncle Benny was fun. I would get the next cow as he would throw in the grain. I would prepare the cow and he would finish the last cow and change the milkers. We worked very efficiently together. He always liked to think of ways to make us more efficient. While the cows were milking we would talk about everything. Our favorite discussion was our invention of a steam car. The greatest dilemma was to reduce the weight. Uncle Benny thought that a propane burner would be the best answer. I would sketch cars on paper towels and we would dream about our super car.

One day I asked why he wasn't married. This was a very touchy subject. He wanted to be married. He spent all his time working on the farm and just didn't get to meet many women. There was one woman that he liked and I asked about her. Oh, he said that was just sour grapes. I didn't know what that meant so he told me the story about the fox that tried and tried to get the grapes but eventually gave up and said, " I'm sure they are sour any way."" Oh, I get it," I said ", You tried but she thinks your too much like an Uncle. On his thirtieth birthday, we went to our favorite swimming hole. We did this every day during the hottest time of the day. He picked up a walking stick and walked down the beach. As he walked, he smoked his pipe. He looked a hundred and thirty. He was very depressed because he was thirty and single.

Uncle benny was cheap. I used to think we must have some Jewish in our family and he got all of it. He smoked a pipe because the tobacco was cheaper than cigarettes. One morning he was shaving and I had to use the toilet badly. As I sat on the pot, I cleaned my glasses with a piece of toilet paper then used another piece when I had finished my duty. Uncle Benny couldn't believe it. "Such a waste! You could have easily reused the first piece. No wonder we have to buy so much toilet paper! And, by the way, You don't need so much at a time". Uncle "Cheap" Benny!

We worked very hard and we had fun doing it. Sometimes we got time off. The work was seven days a week usually. One night there was a special at the local drive in. We went. We didn't eat popcorn. Instead we ate Cheddar cheese and raisins. Whenever I eat cheese and raisins my mind takes me back to that warm summer evening with the smell of cut hay and the fun of being out with my best friend, Uncle Benny.

I always had a best friend in my life. When I went into the army I had many friends. All through training and into Vietnam, there was someone. When I got to Vietnam, it was very hard to have a friend of any kind. Everyone tried to stay distant. We were around each other so much that staying distant was impossible. Many fights started because we couldn't get away from each other, ever. By Christmas I had been in Vietnam about 5 months.

I had friends but typically kept my distance. Joe was assigned to my track. I couldn't believe it. Joe was a jerk. He was very fast to insult everyone and he displayed a special dislike for me. This was not good. For the first two weeks I tried to keep away from him. Joe was assigned to our track because he was the best driver in the platoon and we were point track, I sat on the left machine gun. The left is better then the right because the driver is on the left. A good machine gunner doesn't get excited in a battle and shoot the driver. This sounds simple, but months later, when I was a driver, I was almost shot.

Mostly out of boredom, I asked Joe to teach me to be a driver. Of course, all training for Vietnam came in the field. The army sent us to the war with the worst training I could imagine. Joe was pleased that I recognized his skill. I found him to be very disciplined and a warm person. He had been placed in military school at a young age and had been in the military all his life. At the end of February, he would go home and be out of the army for the first time in his life. He was writing to a girl back home. He didn't call her his girlfriend but he had great hopes. When we discussed women, his experience was very limited. We became good friends. Good enough that, when he went to Australia for R and R, he brought me a present. He did such a good job teaching me to be a driver, by the first of January I was moved to another track when their driver went home.

When I remember Joe, I think of the monsoon season and I remember the Typhoon. Everyday, during this period of time, we would go to the beach and sit. At night we would go to the rice paddys and play war. During the day we sat on the beach. The south China Sea is not like the Pacific Ocean It has waves but the are very shallow. The sea is very salty and warm, about 80 degrees. It didn't take many days and we became tired of swimming. Then we just sat.

One morning we were returning from our night patrol and the waves were very high. We couldn't believe it. Every night it rained during the monsoon season and every morning the sun would come out and it would get very hot. We would drag out our wet cloths and wet bedding and hang it in the sun. It would dry in minutes . Every morning as we drove on the beach we would drive in a dense fog. The fog seemed to be formed from the steam rising from the wet sand. The fog stayed about 10 feet above the ground. The sand at the beach is very white. The whole picture is like a scene from another planet. Five green tracked vehicles driving on white sand moving through a dense white fog. Star wars!

We didn't know why the waves were so high but we were real excited. This morning we didn't stop to dry our cloths. Joe and I grabbed the air mattress and ran into the water. We paddled over three enormous waves and then turned the air mattress toward the beach. We were side by side with the air mattress under our chests. We kicked hard to catch the first wave.

We were picked up and started racing toward the beach. We were surfing! I could almost hear beach boy music in my mind. But, something was wrong. We rapidly moved to the front edge of the wave and started over the top. I looked down from the crest of the wave. Below, about 20 feet was dry sand. I looked at Joe and we went over the top. I landed on the sand head first, still holding on to the air mattress. I hit so hard I thought I had broken my neck. Then the wave hit me and I was tumbled over and over. The air mattress was ripped from my arms. I was then deposited on the beach. I had no idea which way was up. I was dizzy. I staggered on to the shore. My neck hurt. I sat on the beach and held my neck in my hands. Joe walked out of the water and was holding his neck also. I looked to my left and sitting on the beach was the other 15 men in our platoon. They were all sitting on the beach looking at the water. All were holding their necks.

That night we went back to the rice paddies. The army sent us to certain locations to sit. I realized they always had us go to the same spot so the enemy could easily find us. We were sitting ducks waiting for an ambush. When the enemy did hit he was always very ready. To be honest, I resented the way the army put us in this situation. At the time I could only do my best to survive. That night the wind blew very hard. We sealed the track and sent Joe out to watch. The wind and the rain got worse. Suddenly Joe came in the door. He said it was bad. My guard was always the last, from 3:00 am until 6:00. Everyone else switched places but I always took the last because I liked to watch the sun come up. 22 years later, I still wake up at 3 o'clock and go on guard. I told them I wasn't going to sit without a guard. I put on rubber rain gear. Then I put two bullet proof vests over the rain gear. Then I put a poncho over the vests and added one more vest. I put a steel helmet on my head.

The wind blew so hard that I had to sit on my rifle to keep it from blowing out of my hands. I held my glasses on with one of my hands. The wind came from all directions and blew so hard it rocked the 15 ton armored personnel carrier. The most amazing thing was the rain. It felt like someone had thrown a bucket of water. It was so dense that I felt like I was drowning. I had to hold my hand in front of my face so I could breath. I was immediately wet. The water soaked me to the skin before I could get to my station and sit down. I sat there for about half an hour and finally went back into the track. They made me take off all my cloths outside. The cloths just blew away as I took them off.

Inside the track, they had set up the beds. There were five of us so we hung three hammocks from the ceiling and two people slept on the floor. Imagine five people trying to sleep in the back of a Volkswagen bus half filled with boxes. I put on dry cloths and climbed into my hammock.

In my mind I pictured the Viet Cong, hiding in the hills going through the same storm. In my mind, they couldn't get out of the rain. They were very wet and very uncomfortable. One hot headed soldier is angry. Angry that he can't be with his family. Angry that he has to sit in the storm. Angry that he has lost So many friends. What the hell, he is wet anyway. So he grabs his rocket launcher and walks to the place where the Americans always stay. It is raining so hard he is safe. He fires his rocket into the track and none of the Americans are even found until after the storm. As he walks away he remarks that the Americans were very stupid not to have some one on guard. I didn't sleep all night but listened to the wind and waited. I was very relieved when the storm calmed down so I could go out and sit on guard.

After the storm, we went back to the beach to sit. We dried everything and listened to the radio. Every morning we listened to a woman disk jocky. She said, "Hay, how about that typhoon last night. You soldiers in the central highlands really got it last night! That was the first time we had heard about a typhoon. There was a huge gap between the weather people and the soldier in the field. It just isn't like T.V.

I told everyone I was going for a walk up the beach. They were used to that. I hated to sit. I walked about 100 yards and I found a pile of human bones that had been exposed by the storm. The bones had a small amount of green cloth that indicated they had belonged to a solder. I assumed they belonged to a Vietnamese solder that died on the beach in some long ago battle. He had been shot through the head. Maybe even an execution, As I looked the bones, I imagined all kinds of ways he had died. I started back and almost stepped on a second set of bones. This person had been shot in the jaw. The other set were in a pile. This set was perfect. He was laying on his back. The country is covered with bodies. These people have been at war for generations. Everyone at home screams for the remains of our MIA's. I always tell people that you can't apply the rules and experiences of America to Vietnam. It is a different place. Joe was very excited about my find. He took the skull from one body and the jaw from the other body and wired them together. Then he mounted it on the front of our track. He talked to the skull. As long as that skull was on the front of the track, we had bad luck. One day I threw it off the track and into a ditch, Joe was mad but I didn't care.

Joe saved my life. In Vietnam there were many ways to die. After the big storm the valley flooded. We were sitting on the top of a hill looking at the huge lake that had once been the rice paddies. Joe and I were sitting next to each other talking. I was heating a can of "C" rations. Without moving fast, Joe reached into the back of the track and pulled out a machete. Without saying a thing he swung the huge knife down and almost cut off my toes. I wasn't wearing shoes at the time. I almost jumped out of my skin. Then I looked down and almost jumped out of my skin again. Joe had cut off the head of a bamboo viper. This is a foot long green snake that we called a three stepper. He was called a three stepper because he was so poisonous that you took three steps after he bit you and you died. He had crawled out of the flooded valley and was crawling between my feet.

Joe was a great friend and we had many adventures. He loved to "rap". We would sit on the front of the track for hours and talk about home. We even talked about living as room mates in California. I had my grandmother in San Francisco send a newspaper and we spent hours looking at the prices of apartments.

January 31, 1969 Joe Rush was killed. I spent many months hurting silently inside about his death. Such a waste of a good person. I didn't want friends. Friends die.

George Alvin Tyler the Third was assigned to my track early in March. George was big, strong and black. He had an incredible sense of humor. George tried hard to be my friend and didn't even get too upset when I told him I didn't want a nigger as a friend. George and I seemed to always find ourselves in one situation after another. Slowly, we became friends. We spent many hours talking. He enjoyed life and laughed at everything. Women were for love making. That was their only purpose and looks or personality didn't matter. George was very good looking. He would rather seduce an unattractive bar maid then spend money on the most beautiful prostitute. He drank too much, played too much and got into too much trouble. Eventually he and I were best friends. When I talk to people about this period of my life, I always start out by saying, "I used to be black". All the people I ran around with were black. Because George was accepted, so was I. I found out what it means to be a "brother".

One night, we were trying to sneak into a restricted area. This was a wonderful town inhabited by women. It was on the beach of the South China Sea. It was a wonderful place. The warm wind always blew from the sea and it smelled like salt water. We tried to hide in the back of a step van. When we got to the first quard station, the guard opened the back. He was a "brother" and he let us go. The second guard was white and he pulled a gun. George got out of the van and just walked away. The guard kept yelling that he was going to shoot. Later George said that no one will shoot you unless you run. I talked my way out of the problem. George said that I got away with that because I was white. There was an advantage to being a white, black brother.

Today we call sharing things with a friend, Male Bonding. Today's liberated women love to make fun of "male bonding". Being in a war with someone does give a person a chance to Bond. Especially combat. One of our jobs was convoy patrol. We would spread ourselves among a long line of trucks and guard the trucks as they carried supplies to back areas. The Vietnamese would see a convoy and follow. This was their way of getting to remote villages. We ended up guarding long lines a civilians and military vehicles. One morning we were supposed to meet with a group of Vietnamese trucks. As we arrived, I noticed the people had a strange look. It is a look that I saw often and have seen back home after a bad car accident. I was on the left machine gun. I had been training George to drive. When the track commander would say turn left or right, George would salute. I told him that he didn't have to salute when he was driving. He said that was the only way he could remember his left from his right. He was so funny.

We were lead track. I told the track commander that something was wrong. I loaded my machine gun. Then I noticed all the trucks had bullet holes in their windshields. Before I could say anything we were under fire. The enemy was shooting from a hill about 300 yards away. The bullets washed over us. We were firing our guns at a hill! It just went on and on. George yelled that he was hit. I ducked down to help him and saw that he was bleeding from his right hand and his left leg. A bullet had bounced off the front gun shield and went through his hand and into his leg. We didn't wear shirts and we were all very tan. The black men also got darker from the sun. George was so black, he looked purple. It was hot and his blood was flowing fast. I was amazed at how red his blood looked against his purple skin. He was very scared and looked like he may go into shock.

I pulled him back into the track and cut off the leg of his pants. I said ", George, you are a lucky man. The bullet didn't hit any bones and you are going to have a great scar to show the women. You can show them the scar on your hand and then show them the scar on your thigh. What a great way to seduce the ladies. He started laughing. And I said, "look at this, The bullet was sitting under his skin on the bottom side of his leg. "This will make a great item to hang from a chain around your neck. George wanted me to cut it out right there because he was afraid the doctors would throw it away when he got to the hospital. The hospital! All he could talk about was the nurses! Man, this was his lucky day! As I loaded him on the chopper, he was laughing. He got to go for a ride on the chopper and I had to stay and get my head shot off. And he was sure there was a nurse just waiting for him at the hospital.

I returned to my vehicle. The shooting had increased. We had to carry George quite a distance on a stretcher to find a safe place for a chopper to land. I got into the drivers seat and started firing my weapon. Suddenly there was a "woomph" about 100 yards from us. They were firing rockets. The second one came in about 50 yards in front of us. I got on the intercom to the track commander and said to fire the 50-caliber machine gun at the two rounds and walk up the hill with the bullets. It made sense that the rounds were landing between the enemy and us. A straight line up the hill would find the enemy. The next round landed in front of us a few yards. I put the track in reverse and backed up. The next round landed where we had been sitting. This was not fun. If one of those rounds land on us we would all be dead. Woomph!! The explosion threw dirt all over us. I was backing the track up fast. The track commander was giving me instructions so I wouldn't go into a ditch or hit a tree. Then the entire hillside exploded and immediately the bullets and the rockets stopped. I was pissed. I was extremely pissed.

Finally someone had decided to call in some artillery. The artillery rounds kept hitting the hill side. We had sat out there all day. George had been wounded and finally they had decided to call in artillery. Why had they waited? The same answer as always.

Our goal in the war was to get body count. They were hoping we would kill the enemy with our machine guns. You don't get body counts from artillery. There isn't enough left. It was a stupid war!!!

When the lieutenant drove by he was amazed at the number of bullet holes in our vehicle. He said that they hadn't received any rounds. Then he said that we had to go up the hill to see how many of the enemy were killed. My track commander was Bob Hartwell. He was 22 years old and about 6'-4". He was from Boston and claimed to have been in the mafia. He looked at the officer and said, "No, we are not going up there to count bodies! Do it yourself, Sir". The officer got red in the face and walked away.

When George got back in the field, He had some real funny stories about the hospital. He had found a black nurse that was very very ugly. He had seduced her and as George said, "She appreciated the affection of a good looking southern gentleman."

When I left Vietnam I left George. We gave each other our home addresses but he didn't write. Neither did I.

In 1987 I went to Washington D.C. One of the most important names I tried to find at The Vietnam Memorial was George Alvin Tyler the Third. I looked very hard and was pleased to find that his name was not on the list.

I thought for awhile that I was too old for a best friend. I have many friends but something seems to happen when they start getting too close. Perhaps, life moves so fast that it is difficult to have best friends.

I went to Vietnam when I was 21. I have been out of Vietnam 22 years. Perhaps is time to let go. Perhaps I expect too much. I only ask one thing of a friend. That they promise never to die.

CHAPTER 16

COMING AND GOING

Finally, School was out! I was in Junior High School and was waiting for the final bus to take me home. The next day I would get on a greyhound and go to the farm. It had been a tough year. I had flunked one class, I had gotten into a fight and had been knocked out, I had been cut off the basketball team and the football team and I had been rejected by the best girls in school. No it had not been a good year.

I couldn't wait to get out of Seattle. The next day I loaded onto the bus and headed for Eastern Washington. The sun was shining and I settled into the long trip with a spiderman comic book. Suddenly a boy my age sat next to me. He introduced himself and told me he was going to his grandparents home in Spokane. He was a little overweight and not a real good looking kid. In Junior High School, good looks was the first item on the list for judging a person. His excess weight said that he wasn't a good fighter. This was the second way of judging someone. He was definitely a loser. And he wouldn't stop talking! This could be a long trip.

Without even stopping to see how I felt, he turned to two girls across the aisle from us and introduced himself and then me.

One of the girls was cute and the other was just ok. He started to flirt with them! Then they started to flirt back! Finally he asked me to change places with the cute girl. I moved over and the almost cute girl started to talk to me. I was very uncomfortable.

After we talked about where I was going and where she was going, I didn't know what to say. I looked at my new friend and he was talking away. I turned to the girl next to me and said, "Look at him talk! What does he have to say? I think he is kind of a jerk but he is making your friend laugh. Maybe you would like me to sit somewhere else."

She wasn't real pretty but she looked me right in the eyes and said," NO I don't want you to move. I like you. I'm a little uncomfortable too but this is fun. I thought the trip would be a real bore. Lets have fun and enjoy ourselves." Then she yelled across the aisle that my friend sure talked a lot. She started teasing him about having such a big mouth. He just kept talking and then he asked if I was trying to be the strong silent type. If I was, I only had it half right. At the time I was just under six feet tall but weighed about 120 pounds. This made everyone laugh but me. I blushed. I was very self-conscious about my body. This made them laugh and my "girlfriend" said I was cute when I blushed. This made me blush in a different way.

We played and laughed and flirted the entire six hour bus trip. When we arrived in Spokane, my friend talked the two girls into not calling their relatives for awhile. We went into the restaurant at the bus stop and had a coke. I was real glad I hadn't spent all my money on spiderman comic books. I had to pick up half the tab.

After a while all three of them made their phone calls and suddenly, I was alone. I caught my bus and headed for Kettlefalls. The closer I got to the farm, the better I felt. I sat in the front seat of the bus and talked to the driver the entire trip. There was only one other person on the bus. We talked about hauling hay and farming and working hard. He had owned a farm before he was a bus driver.

Before I knew it, we pulled into Kettlefalls. As we drove down the center of town I recognized the little Catholic Church that Grandma had been commissioned to paint. I had thought the new church would be finished by now. Then past the movie theater. Thunder Road was playing. I had already seen it. The bus stopped and I got out.

I was hit with the smell of alfalfa and the heat of the sun. It was getting late but the sun was hot. Everything was dry. I felt good.

No one picked me up. I couldn't call because there wasn't a phone at the farm. I sat on my suit case and looked at the hill behind the town. It looked hot and dry and covered with sage brush. I wondered how it would look like in one of Grandmas paintings. Everything felt good and free. I felt like a trapped animal just let out of a cage. I never wanted to go back to Seattle. Never!

Finally, I went to a gas station next of the bus station and bought a candy bar. Just as I got back to my suitcase, Uncle Danny drove up in his red international pick up. He was very happy to see me. I was very happy to see him.

As we drove up the dusty road to the farm, I could smell the pine trees and feel the space and openness of the farm. When we walked into the house, Uncle Benny was sitting at the table next to the door. He was smoking his pipe and drinking a cup of coffee. I felt a little strange when he shook my hand but I enjoyed the feeling that these people were really glad to see me.

The next morning, my first job was to get on the new International Harvester tractor and scoop the Cow poop out of the waiting area in front of the barn. This was a big deal because this tractor was diesel. I had to load all the poop into the Manure spreader. Uncle Benny showed me how to run the tractor, then left. Man, was I feeling like a big shot! At home I felt like I couldn't do anything. Here they let me drive the diesel tractor! The first thing I did was drop the scoop and move the tractor forward, just like Uncle Benny told me. Then I grabbed the hydraulic levers and picked up the scoop. There was a groaning sound and the back wheels of the tractor came off the ground. I immediately pushed the levers forward. I had pushed the scoop under the corner of the barn. My heart sunk. My first job and I had destroyed the barn and the tractor. I got off the tractor to look at the damage. It looked all right So I got back on the tractor and went back to work.

About the time I finished loading the poop into the manure spreader, Uncle Benny came around the barn. He asked me how it was going and did I have any trouble. I answered that everything had gone very smoothly. What did he want me to do next? He said, of course, I had to go out into field number one and spread the manure. Uncle Benny said "Shit" instead of poop. When I get older I'll say shit too.

We hooked up the Manure Spreader to the tractor. This piece of equipment looked like a metal wagon with wheels at the back. The wheels had long teeth and, when thrown into gear, the wheels would turn and throw the poop out the back. I had never run this piece of equipment before. First day was a lot of fun.

When I got to the field, I thought I would impress Uncle Benny by getting the work done fast. First thing was to take off my shirt. It was starting to be a hot day. Next, I threw the Manure spreader in gear and the tractor into third and high. I took off like a shot. What I didn't know was that the manure spreader threw poop in a wide arc. The faster it went, the farther the poop was thrown backwards. Also the higher it was thrown into the air. Because the cow poop was fresh, it was runny. Also the faster the tractor went, the farther forward the cow poop was thrown. When the tractor started, I was looking forward. I was very determined I would get the job done well. The first lump of fresh cow poop hit me in the back of the head. Then my bare back was covered. Instead of stopping the tractor, I turned to see what had hit me. A big lump of wet green cow poop hit me in the corner of my mouth, I stopped the tractor and started laughing. I laughed so hard I was rolling on the ground. I finished the job at a slower pace. Yes, I loved the farm. It was a place where people trusted me and a place where I could make mistakes.

At the end of the Summer I was tan and full of confidence and I didn't want to go home. I was going to ride with Uncle Jay to Wenatchee where he lived. Then my mother would pick me up. My cousin, Sharon, was spending the summer baby sitting for Uncle Jay. Uncle Jay is a dentist. Sharon baby sat so she could get her teeth fixed. She was 15 at the time.

As we left the farm, I felt very sad. Uncle Jay kept telling me how proud he was that I had done a mans work. He was fun and kept cracking dumb jokes. His humor was silly.

Suddenly, he stopped the car. "Look, a crop duster". We watched the crop duster from the side of the road. It flew right over our heads. I was amazed at how low he flew. Uncle Jay liked to see him fly next to the power lines and the trees. He seemed to be waiting for a crash. When the crash didn't happen we loaded back into the car and were off.

These were the early days of being a Dentist, when much of his money went into the practice. He lived in a new house on a hill. My brother and I were amazed whenever we stayed at this house because the builders had forgotten to finish the basement.

When we drove up to the house, Uncle Jay suddenly got real mad. Aunt Joan had bought something for the front of the house. He was going to "talk to her".

I didn't wait around to see how this turned out. I shot through the house and into the back yard. Sharon was lounging in a lawn chair getting a tan. She was a shock to me. Sharon had been "just" a cousin for a long time. Some strange things had happened. She got up and was wearing a pink bikini. I felt like I could hardly breath. She smiled and said that I was going to catch flies if I didn't close my mouth. I closed my mouth so fast my teeth clicked. I said," Sharon, you are beautiful!". She said, I know". and laughed.

About that time my mother drove up from Seattle. Everyone was real happy to see each other. Finally Sharon and I were alone again. She told me that Uncle Jay had made her work all summer and wouldn't let her meet any boys. She called him a horse doctor, and an old fart. I told her that I liked Uncle Jay and that she shouldn't say fart. She laughed at me. I could talk to her now because she had put on a tee shirt over her bikini. She was the same old Sharon.

We were going to have dinner at Aunt Joans parents house. I was real glad to hear that. Her mother was a great cook. When we got there the men went out to the back yard. I went along. Uncle Jay kept teasing his father in law. He would joke around about something I didn't get and the older man would just smile. It was easy to see that they both liked each other and saw each other often.

From a concealed place , Uncle Jays father-in-law pulled out a bottle of his favorite whiskey. He took a big drink from the bottle and handed it to Uncle Jay. He took a big drink and handed it back. His Father-in-Law asked if it was ok for me to have a drink. Uncle Jay didn't seem to really like the idea but answered that I had worked like a man all Summer so it was OK. I took the bottle and tried to take a long drink like the men. When the liquid hit my throat, I stopped drinking. I wanted to choke. I really wanted to choke but I wasn't going to. I couldn't. Not in front of the Men. I handed the bottle to Uncle Jay. He smiled. At least I think he smiled. My eyes had welled up with tears and I was having trouble seeing. But I wasn't going to choke. Uncle Jay took some, gave it to His Father-inlaw and He Gave It Back to Me! I took another drink. I was hopping it would stop the need to choke. This time it was a little bit better. In fact, I felt quite grown up and quite good. We heard a call from the house that dinner was ready. I could smell roast and gravy. My mouth watered. I had a little trouble getting up the back stairs. I lifted my foot high enough but it hit the top stair anyway. I felt good.

As we sat down to eat, I sat next to Sharon. We got all settled in and Aunt Joan asked Sharon what she wanted to drink. Sharon said "milk". Then they turned to me. I smiled confidently and said, "wine". I looked at Uncle Jay and he was trying to keep from laughing. I was having a great time. Aunt Joan and my mother both said the wine was for grown ups but Uncle Jay, with his most masculine voice, declared that I had worked like a man and should have wine. I got a glass of Apple wine. It was very tart. I had three glasses. Uncle Jay poured all three. I didn't care what Sharon said, I liked him. On the third glass, Sharon kicked me under the table. She was two years older then me but she had to drink milk. I slept on the drive back to Uncle Jays house.

When my mother, Sharon and I got back to Seattle, my mother asked if Sharon and I would like to go to the movies. Of course, we said yes. My mother gave the money to Sharon because she was the oldest.

At the movies, there were a lot of teenagers waiting in line. Sharon had a summer dress on and was gorgeous. I felt like a king. She would lean close to me and we made fun of all the other people in line. I would make up some strange story about where they came from and she would laugh. She even put her hand on my shoulder. I felt like I was on a date. We were having a great time. I had brought along all my money so we could have candy and popcorn. By the time we got to the ticket counter, I had convinced myself that every person in line thought that I was on a date with this very beautiful girl. I reached into my pocket the way any male would on a date and pulled out my money. The ticket lady looked at me and said, "I'm sorry sir, but you are two dollars short." Sharon reached into her purse and pulled out the money my mother had given to her and paid for the tickets. I wanted to die. Yep, I was back in Seattle!

Years later I was on my way back from Georgia. I had just completed paratrooper school and was on my way to Vietnam. I had a thirty day leave. I was anxious to get home. I had been in the army for six months and I hated it. Besides, I couldn't wait to see my girlfriend.

The last day before I went into the army. She had called me over to her house. She and I had been very careful. I wanted to marry her when I got out of the army. We had even spent many evenings naming our children. The boy was going to be Kenneth David and the girl was going to be Heather Dawn. We were careful because we both wanted her to be a virgin when we got married. When I got to her house, she had told me there was something she thought was important. The house was empty. She took me upstairs to her parents bed room and took off all her clothes and all of my clothes. She gave me a big kiss and suddenly the room was full of light. She yelled that her parents were home, grabbed her clothes and ran into the bathroom. I hurriedly put on my clothes and, just as I started to tie my shoe, The door opened. It was her older brother. "What were you doing with my sister?". "I don't know what you are talking about", I answered. He said, "oh yey, What is that ? He pointed at her bra sitting under my left shoe. I Just looked at him. He went to the bathroom door and told his sister to come out. She said, "no". He hung her bra on the door knob and left.

After she came out, I left too. Now I was going home. I was going to go to war and if she was that way when I went into the army, think how wonderful she would be now that I was going to war!

The first night, I took her to Chinese dinner. During dinner, she explained that she was not going to wait around for a year for me to come home, if I did come home at all. She said that, if I were still alive after I got out of Vietnam, I should give her a call,

So, I spent most of my thirty day leave feeling like shit!

A few days before I went to war, My mother had a party for me. Most of my Uncles, Aunts and Cousins came. Uncle Benny even got me a date. We started the party with mai tais. We were having a great time. Sharon, Pam and Linda were there. These three are all sisters and are the most beautiful women in the world. After a few drinks, Sharon bumped into me in the kitchen. She gave me a kiss. Then she opened her mouth. I was amazed. When we were done kissing, she called Pam. "Hey Pam, give Steve a sexy kiss. He is pretty good". Then they took turns kissing me. The mai tais and the kissing made me dizzy.

Suddenly my date walked up to me. She had been drinking mai tais also. I had been having such a great time with my cousins that I completely forgot about her. She said I was a stupid shit. Then her eyes rolled back and she slowly slid down the wall. Uncle Benny ran over to her and kept saying over and over, "Her mother is going to kill me." He got her to the bedroom and laid her down just as she started throwing up all over. It smelled like pineapple. When my leave was over, I stayed at Fort Lewis waiting for a plane to take me to War. Day after day I waited. They would call us out for formations and then we would spend the day waiting. After about a week I was getting to be good friends with a soldier from South Carolina. I turned to him and asked if he would like to see Seattle. He said sure. We walked over to the bus station and got on a bus. He turned to me and said, Do you realize we are going A.W.O.L.? I answered sure, but what could they do to us if they caught us, send us to Vietnam? He answered that he heard they could shoot us. Oh Well.....

In Seattle, we walked through the Seattle Center and into The Black Angus Restaurant where my mother worked. She was working so we went into the bar. I was 20 and my friend was 18. My mom told everybody that we were going to Vietnam so everyone bought us drinks. I told my mom that the plane had been delayed so we weren't going for a couple days. When she drove us home my friend threw up on the side of her car. The next morning, we jumped up and Mom drove us to Fort Lewis. We arrived to see everyone lined up. They were calling attendance. As I got into formation, they called my name. Then they said that everyone that had been called had to get on the bus. The bus took us to an airplane and we were off to war.

As I flew into Vietnam, I saw tracer bullets on the hillsides around the airport. I said to myself, "This is where I am going to die". I didn't know that the army shoots lots of bullets into the bushes just in case someone is there. Especially when an airplane is landing.

When we landed, I couldn't believe the heat. It was so hot I almost couldn't stand it. As we walked off the runway, there was a plane next to ours. It was loading aluminum boxes into the back. One of the soldiers asked what was in the boxes. I had already guessed. Those, the sergeant answered," are the men you are replacing". I felt sick.

I started counting days when I had 300 to go. Each day I marked it off my calendar. Finally, I had 30 days left and I had to make a decision. If I stayed 30 more days beyond a year, I would be out of the Army. The army believed that if you had 5 months left, it was too much trouble to train a soldier. If you had 6 months, you were worth the money they spent on you. I wanted out of the army. I signed up for 31 more days. I was very afraid I would die in the last days. As I look back on that time I hate the people that made me make that decision. Could there be any thing more cruel?

17 days had past when I could have gone home. I was laying on a cot under a tree. It was hot. Bullets were being fired. I rolled off the cot and ran to my machine gun. Later I went back to the cot. The bullets had hit the tree less then 6 inches above the cot.

We had been given orders that we could not shoot unless we saw the enemy. I was looking at a wall of elephant grass. This is so thick and tall only an elephant can get through it. The bullets kept washing over us. I loaded my machine gun and fired. 100 rounds into the grass. Immediately the shooting stopped. About that time the captain arrived. He jumped off his vehicle and asked who had fired. I answered that I had fired. He asked if I had seen the enemy. I said, "no". His look was not good. Immediately I answered that I had seen the smoke from his rifle. He asked where and I pointed into the grass. We drove into the grass.

As it turned out, it was only about 100 feet deep. On the other side was a palm tree grove. Every tree had the bark ripped off about chest high. When I looked closer I could see where my bullets had ripped into the trees. The captain was very unhappy that we didn't find blood. My final days of the war were insane. Finally 7 days left. A chopper picked me up and took me to Ahn

Ke. This was the 173rd headquarters. It was not impressive. They took my rifle away. This was the first time in over a year I was away from my rifle. This weapon had saved my life. They took it away and told me to rely on the others for protection. I came home because I took care of me.

The first night I had terrible nightmares about the enemy overrunning the camp. By noon the next morning I began to drink Calvert and coke. By dinner I just sat at the end of the bar and felt nothing. I drank until I threw up and then I drank some more. Each night I would pass out and wake up too late for breakfast. I would go to the bar and order french fries and a calvert and coke. For five days that was my life.

Next they sent me to Camron bay. There were hundreds of soldiers waiting to go home. They told us they would call our name once. If we didn't hear our name they would leave us in Vietnam to rot. Names were called out constantly from speakers on telephone poles. They were called out 24 hours a day.

For three days I took turns with one person. We slept in our clothes. One of us would go eat, the other would sit on guard waiting to hear our names. We would take turns sleeping. One on guard for three hours, the other sleeping. On the third day, as the sun was coming up, they called my name. They did not call his. I was on guard at the time and I was not even sure it was my name. I had waited so long that I thought I had dreamed it. I grabbed my stuff and ran to the airplane. For one quick second, I felt sorry for my friend. They hadn't called his name.

The airplane had young attractive stewardesses. I had not bathed for three days. My cloths did not fit and I hadn't slept for so long, I couldn't remember what it was like. I woke up once when we landed in Japan. I bought a bottle of Calvert to remind me of the horror I went through. Before I was in the plane very long,

I and the soldier next to me drank all of it. Then I fell asleep. I woke when someone cheered. I looked out of the window and saw evergreen trees and cars driving along I-5.

I couldn't stop myself. I began to cry.

AFTERWORD

Stephen Manthey passed away on August 19th, 2018.

Stephen Manthey served in the Vietnam War from 1968–1969 in the first platoon in the US Army Company D, 16th Armor/17th Cav Airborne Brigade. Stephen's Company consisted of three platoons of Armored Personnel Carriers, (APCs) or (Tracks). He was a driver on track one-seven and operated the 50-caliber machine gun on the left side of the track. While serving in Vietnam, Stephen earned a Purple Heart, National Defense Medal, an Army Accommodation Medal, and a Republic of Vietnam Campaign Medal with Four Bronze Service Stars. He was a lifetime member of The Sky Soldier 173rd Provisional Tank Company Society of the 173rd Airborne Brigade.

These stories were written by a combat veteran who had struggles with PTSD in his life after serving his country in the Vietnam War. The intent of this book is to join the journey that Stephen Manthey went on to try to find answers that One Year, One Month, and One Day had on his life. By sharing Stephen's stories about his time in the Vietnam War and the effects it had on his life. His son, Brian Manthey, hopes that it will help other Veterans with their journey.